U0422791

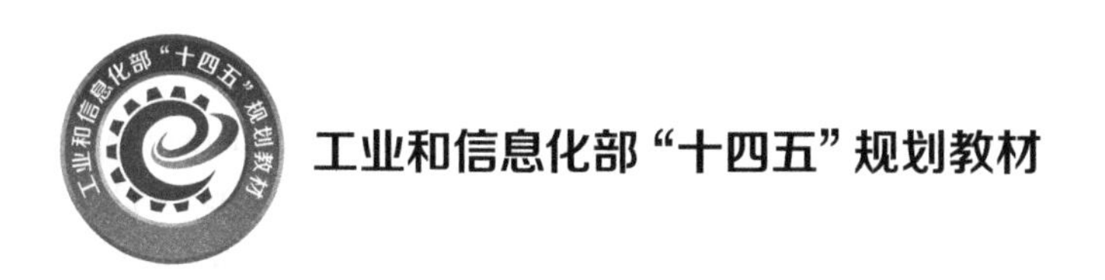

Material Flow Analysis and Its Applications

物质流分析及其应用

主　编　曾现来
副主编　童　昕　陈伟强　巩如英

北京航空航天大学出版社

Abstract

This book is listed as the 14th Five-Year Plan textbook of the Ministry of Industry and Information Technology. In 2020, Tsinghua University started the English course "Material Flow Analysis and its Applications" for the first time in China and made it a global public course. This is the textbook for the English course. The textbook is divided into five chapters, including the introduction, material flow methodology, applications and cases, emerging methods, and the way forward.

The book can be used as a textbook for students majoring in environmental, management, geography, chemical or other related majors, as well as a reference book for engineers and technicians in related fields.

图书在版编目(CIP)数据

物质流分析及其应用 = Material Flow Analysis and Its Applications / 曾现来主编. -- 北京 : 北京航空航天大学出版社, 2023.6

ISBN 978-7-5124-4086-9

Ⅰ. ①物… Ⅱ. ①曾… Ⅲ. ①环境资源—资源管理—研究 Ⅳ. ①X37

中国国家版本馆 CIP 数据核字(2023)第 078030 号

Material Flow Analysis and Its Applications

物质流分析及其应用

主　编　曾现来

副主编　童　昕　陈伟强　巩如英

策划编辑　董　瑞　　责任编辑　龚　雪

*

北京航空航天大学出版社出版发行

北京市海淀区学院路 37 号(邮编 100191)　http://www.buaapress.com.cn

发行部电话:(010)82317024　传真:(010)82328026

读者信箱:goodtextbook@126.com　邮购电话:(010)82316936

北京建宏印刷有限公司印装　各地书店经销

*

开本:787×1 092　1/16　印张:11.25　字数:295 千字

2023 年 6 月第 1 版　2023 年 6 月第 1 次印刷　印数:1 000 册

ISBN 978-7-5124-4086-9　定价:66.00 元

若本书有倒页、脱页、缺页等印装质量问题,请与本社发行部联系调换。联系电话:(010)82317024

Foreword

Material Flow Analysis (MFA) quantifies and evaluates the mass flow and processes of substances in the system over a period of time. The Millennium Ecosystem Assessment methodology is now being linked to environmental input-output assessments, scenario analysis and life cycle assessments, which are increasingly comprehensive assessments that are expected to be central tools for sustainable development of technology-economic system. Since 2010, it has been widely used from academic exploration to government decision-making to industrial technology upgrading.

In 2020, I designed an English course entitled *Material Flow Analysis and Its Applications* at Tsinghua University. It is the first public course namely as material flow analysis in China. The course has absorbed many students who major in solid waste treatment, environmental management, industrial ecology, or environmental system. Additionally, a number of graduates from other schools like school of chemical engineering, school of material, school of architecture, and school of economics and management were also involved in this course. This course was incorporated into Global Open Course at Tsinghua University in 2022. Based on the teaching material, I proposed this textbook, which has been selected as one of the textbooks of the 14th Five-Year Plan of the Ministry of Industry and Information Technology of China.

Handbook of Material Flow Analysis: For Environmental, Resource, and Waste Engineers, developed by Vienna University of Technology, Austria, is the first and classical course in the world. In recent years it has been used as a textbook for our class. However, in 2020, Ministry of Education, China introduced the ideological and political construction regulation for higher education course. The curriculum ideology and philosophy should be immersed into the whole process of course teaching. Secondly, on the 20th Communist Party of China National Congress in March 2023, it was pointed out that education, technology, and talent are the fundamental and strategic support to comprehensively establish a modernized socialist country. This spirit should be well-equipped in the new textbook. Lastly, the origin of MFA is to be found in waste management, since it became necessary to close cycles and reduce the amount of waste. The approach of MFA is now well established in many industries and increasingly popular. Therefore, these three issues inspired me to write this textbook which is suitable for the Chinese education needs.

Again, MFA is an interdisciplinary field. It is related to not only earth science, environmental science, material science, engineering, and ecology, but also management science, economic science, and law. Thus, I invited three experienced scholars to participate in the writing of this textbook: Prof. Xin Tong on social science for Chapter 4, Prof. Weiqiang Chen on earth science for sub-chapter 3.1, and Prof. Ruying Gong on ecology for reviewing all the materials.

I acknowledged the fund support from Graduate Student Education Teaching Reform Project of Tsinghua University (grant number: 202201J008 and DX04_08) for this textbook. I was also cordially grateful to Beihang University Press to fulfill this smooth publication.

Xianlai Zeng

March 7th, 2023

Contents

Chapter 1 Introduction

Three hundred years ago, the great physicist Sir Isaac Newton proposed the Three Laws of Motion. It indicates that all objects are in motion and the flow of matter is common in the earth. The earth is naturally divided on a global scale: the lithosphere, atmosphere, hydrosphere, and biosphere. Driven by the human, the anthroposphere encompasses all human beings throughout the earth system including our culture, technology, built environment, and related activities. Physically, it consists of the cities, villages, energy and transport networks, farms, mines, and ports.

Since the end of the eighteenth century, our earth has entered the era of the Anthropocene (or the Anthropocene Epoch). The human-influenced, or anthropogenic activity, is dramatically altering atmospheric, geologic, hydrologic, biospheric and other earth system processes. As a fundamental approach, natural (or geological) mineral is extracted from the lithosphere, then produced and manufactured into the product to meet the demand of human and function the whole society. During the process, some toxic substances are released to the environment, like hydrosphere for water pollution, atmosphere for air pollution, and pedosphere for soil pollution. Material flow occurs from the underground to aboveground in product or waste, generated from different processes.

Unsustainable exploitation and usage of natural resources is essential to our economic progress. Global material resource extraction has quadrupled over the last 50 years, and it is expected to at least double as the rise of developing nations. We are exploiting and utilizing our planet's resources much faster than we are replenishing them today. On the other hand, a shift to a low-carbon economy and energy system could increase demand for some important minerals and strategic resources. Copper use, for example, will peak between 2030 and 2040. Overall, many geological metal minerals are in short supply or may have been depleted (Figure 1 - 1).

Environmental pollution is also an important issue linked to resource consumption and scarcity. The release of materials and substances into the environment causes pollution. Solid waste has been exploding at an alarming rate. In 2010, the global generation data for municipal solid waste (MSW) and industrial & construction solid waste was 2 billion tons and 7～10 billion tons (1 ton = 1 000 kg), respectively. In 2018, only China generated 10～12 billion tons of solid waste, of which 60～70 billion tons were dumped without being disposed of. This century will be the peak of global waste output. Global demand for environmental measurement is increasing, and newly revised legislation allows for cost internalization.

Traditional linear approach is based on the extraction from natural mines, production, consumption, discarding, and landfilling. With the constantly expanding populations and

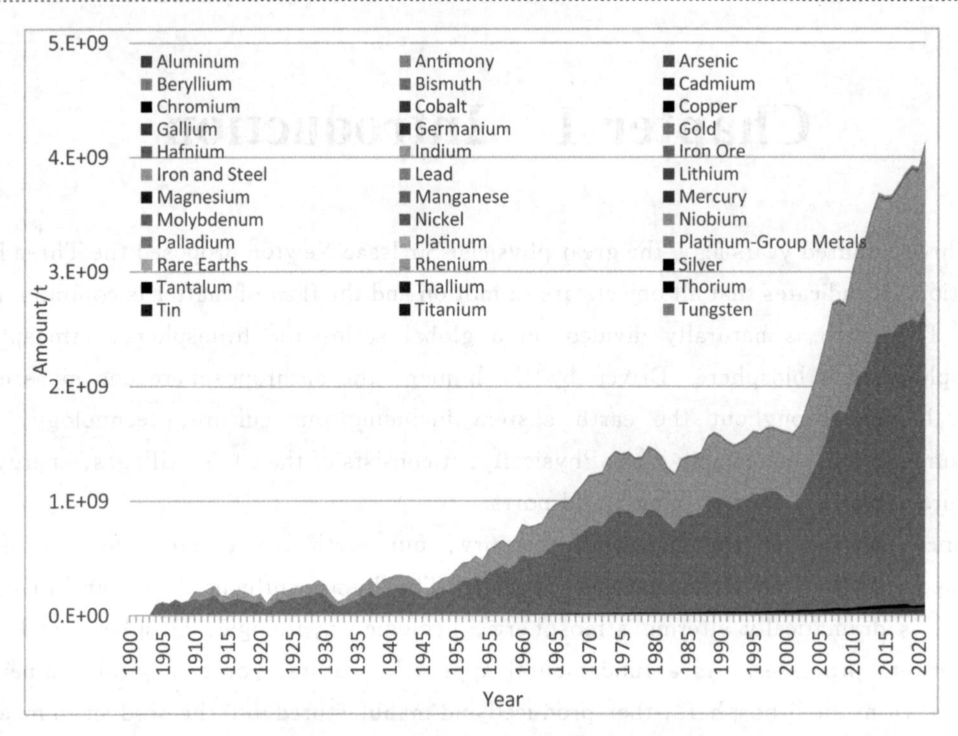

Figure 1 - 1　Global yearly primary mining production of metal mineral commodity (Data source from USGS)

technological innovation, a great deal of resources has been transferred from natural mines to anthropogenic stocks of materials. The shortage of resources drivers waste recycling in the circular approach or closed-loop supply chain (Figure 1 - 2). It creates the base for the development of urban mining concept.

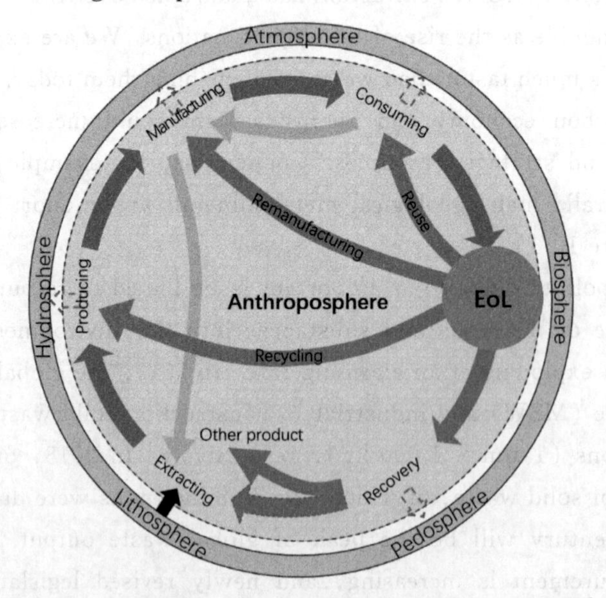

Figure 1 - 2　Material flow for the closed-loop supply chain①

① Reproduced by permission from Zeng X, Li J. Urban Mining and Its Resources Adjustment: Characteristics, Sustainability, and Extraction[J]. SCIENTIA SINICA Terrae, 2018, 48 (3), 288-298.

1.1 Scope and boundary

Since the movement is ubiquitous, the material flow exists from microscopic and mesomorphic to macroscopic scale. The basial boundary for material flow analysis (MFA) is initially defined here (Figure 1 - 3, the dash line circle indicates an unpopular scale for MFA, and the longer dash line circle, the wider use of MFA):

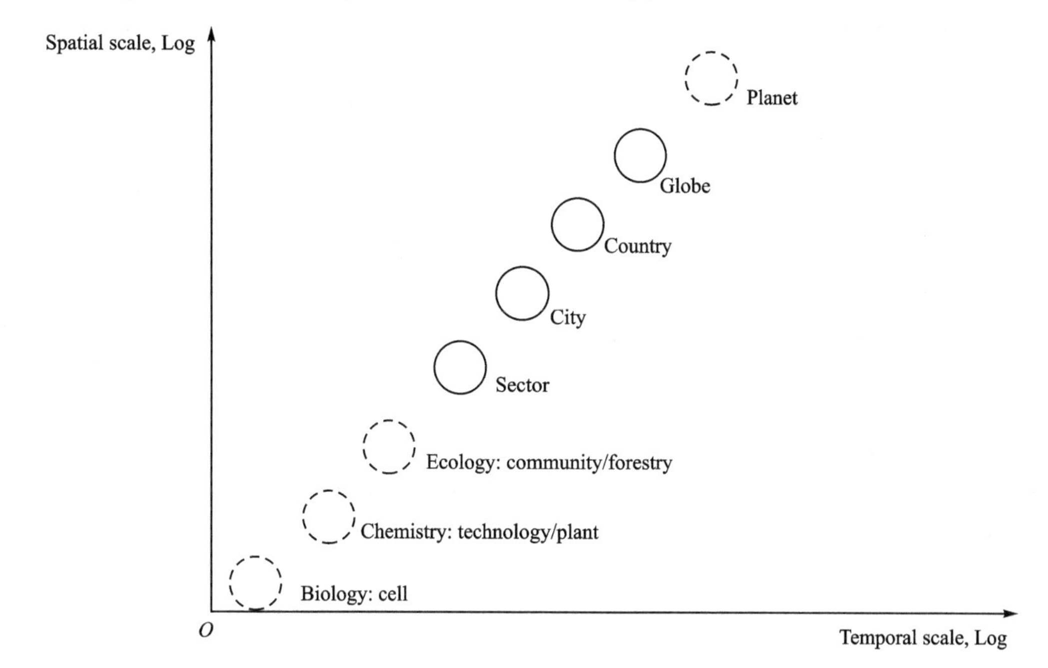

Figure 1 - 3 Scope of material flow analysis in the temporal and spatial scale

Biology: The cell is the basic unit of life discovered by Robert Hooke in 1665. Their scales range from 0.000 1 mm to nearly 150 mm beyond. Driven by the metabolism, they constantly provide structure to the body and convert the nutrients taken from the food into energy.

Chemistry: Reaction is the core that moves the chemical production from technology to factory. Mass transfer and thermal transmission often occur during the reaction process. The reaction time at the temporal scale majorly ranges from few seconds to few minutes.

Ecology: Ecology is the study of organisms and how they interact with their surroundings. The species and the environment are concerned here, very like as community and forestry. Some material, substance, and energy flow between the environment and species. A typical example is that substances can flow along the food chain.

Sector: The industrial and commercial sector is a unique part or branch of a nation's economic or social or of a sphere of industrial activity. In the industrial and commercial sector, product-related materials flow along manufacturing, distribution, consumption, and recycling. The very name industrial ecology conveys some of the content of the field, which can be defined as "the study of material and energy flows in industrial and consumer activities, their impact on the environment, and the influences of economic, political, regulatory, and social factors on the flow, use, and transformation of resources".

City, nation, and the global: Sectors and the environment make up the human habitat. City is the main intensity area of resource consumption to discarding. Urban metabolism, derived from bio metabolism, is a model to facilitate the description and analysis of the flows of the materials and energy within cities, such as undertaken in an MFA of a city. The bigger scales as nation, region, and the globe are also well concerned within the scope of MFA. In fact, most MFA applications fall within these scales (Figure 1 - 3), because they are much closer to our lives.

Planet: The planet is recognized as any of the seven celestial bodies sun, moon, Venus, Jupiter, Mars, Mercury, and Saturn that in ancient belief have motions of their own among the fixed stars. Although natural movement (such as meteorite) or human activity (like spacecraft) occur, there are few considerations on MFA, and thus the planet is the above boundary of MFA application.

From insight of disciplines, MFA using full life cycle thinking is devoted to answer the entire ecosystem problems. It has evolved from a number of multidisciplinary areas (Figure 1 - 4). Environmental science addresses the pollution prevention and controlling caused by toxic substance release. Materials science, the study of the properties of solid materials and how those properties are determined by a material's composition and structure. Green chemistry refers to the design of chemical products and processes that reduce or eliminate the use and formation of hazardous chemicals. Mechanical engineering is the application of the principles and problem-solving techniques of engineering from design to manufacture to the marketplace for any object. Ecotoxicology looks at the impacts of contaminants including pesticides on individuals, populations, natural communities, and ecosystems. MFA has a strong implication for policy making, which is well involved in management science.

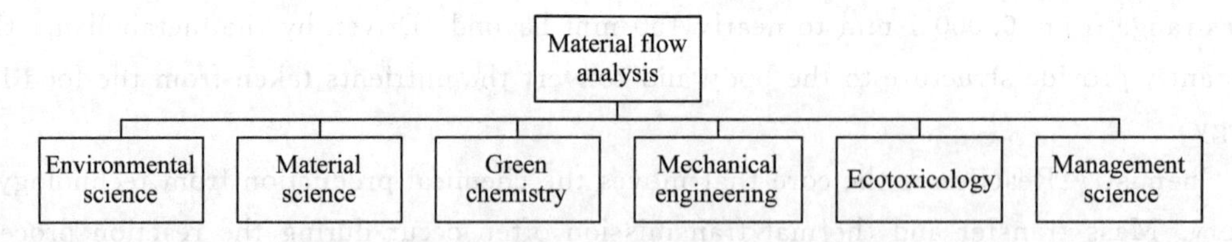

Figure 1 - 4 Major disciplines related to material flow analysis in academic field

1.2 Concept and history of material flow analysis

1.2.1 Concept and classification of MFA

In order to examine the material flow in the boundaries, some analysis is quite necessary as an emerging method. Material flow analysis (also known as material flow accounting) is a systematic and quantitative assessment of the flows and stocks of materials within a system defined in space and time. It includes the comprehensive measurement of the content input and output flows into space at a time specified framework.

Here, material can be replaced by substance, goods, energy, and waste. Substance flow analysis (SFA) is used to measure the flow and stock performance of some chemically pure material (like Cu and CH_4). Energy flow analysis (EFA) is an analytical tool for prediction of the frequency-averaged vibrational response of built-up structures at high audible frequencies. Waste flow analysis is to describe the process of waste generation, collection, transportation, composting, recycling, incineration, and landfill.

MFA has been widely applied to material systems in providing useful information regarding the patterns of resource use and the losses of materials entering the environment. MFA and life cycle assessment (LCA) are traditionally different tools for environmental decision support. The two methods are basically different with respect to the definition of system boundaries and the actual subject of investigation. However, there are also overlaps between the tools. These overlaps highlight that MFA and LCA can complement each other and thereby increase the quality of studies in both domains. The combination of these tools will therefore make it possible to offer the potential for more consistent and reliable decision-making support in environmental and resource management.

1.2.2 History and development of MFA

The use of material flow principle to investigate metabolic systems has a long tradition, dating back 400 years ago to studies by the medical doctor Santorio Santorio, who established an input-output balance in the body's metabolism. In this way, he discovered that around half of the input was not to be found in the outputs (body stock, urine, feces) he measured. Even though he was unaware of the mass of the respiration leaving his body, he knew that there was a missing term in his balance, which he called "insensible perspiration".

Box 1 - 1: Santorio Santorio (1561—1636)

Santorio was Italian physician, known as the father of experimental medicine. He was the first to employ instruments of precision in the practice of medicine and whose studies of basal metabolism introduced quantitative experimental procedure into medical research. Santorio was a graduate of the University of Padua, where he later became professor of medical theory (1611—1624).

Around two centuries later, the famous French chemist Antoine Laurent Lavoisier finally elucidated the law of mass conservation, with mass balances being the foundation of chemical experiments from thereon. Though the principle of mass conservation is fundamental to all kinds of MFA, different MFA types can be distinguished according to the major goals of each type of approach. The historical sugar mill machinery was started from chemical

Box 1 - 2: Antoine Laurent Lavoisier (1743—1794)

Lavoisier was regarded as the father of modern chemistry. He promoted the Chemical Revolution, naming oxygen and helping systematize chemical nomenclature. He established the law of conservation of mass, determined that combustion and respiration are caused by chemical reactions with what he named "oxygen", and helped systematize chemical nomenclature, among many other accomplishments.

material reaction (Figure 1 – 5). However, in practice no sharp borders exist between these types and many characteristics are common to all types of MFA. An important distinction nevertheless can be made between black box-type material balances and refined material flow analyses distinguishing various elements within the system. The former is often referred to as "material flow accounting", where the system under investigation is treated as a black box, for which material inputs and outputs are balanced. The latter is typically referred to as "substance flow analysis" or "material flow analysis", building on material flow models to describe the pattern of material use in the system under study in detail. In this section, the latter type of MFA studies is in the focus.

Figure 1 – 5 Sugar mill machinery, historical artwork①

The production system is described as a network of flows of goods (provisions) between the various production sectors. Input-output analysis of economic sectors has become a widespread tool in economic policy making. It proved to be highly useful for forecasting and planning in market economies as well as in centrally planned economies, and was often applied to analyze the sudden and large changes in economies due to restructuring. In this field, Wassily Leontief seeked analytical tools to investigate the economic transactions between the various sectors of an economy.

> **Box 1 – 3: Wassily W. Leontief (1906—1999)**
>
> Leontief was an American economist of Russian origin. His research was focused on the interdependence of anthropogenic production systems. His research was focused on the interdependence of anthropogenic production systems. One of his major achievements, for which he was awarded the 1973 Nobel Prize in Economic Sciences, was the development of the input-output method in the 1930s.

① Sugar mill machinery on display, historical artwork. This is the International Exhibition of 1862, held from May to October of that year, in London, UK. It was also called the Great London Exposition. This sugar mill was provided by Mirrlees and Tait of Glasgow, Scotland. A flywheel is at right, driving a set of cogs and gears, which in turn operate the grinding part of the mill (center left). On the upper tier are large boilers. This exhibition had 28,000 exhibitors from 36 countries, representing industry, technology, and the arts. There were some 6.1 million visitors. Photo by GEORGE BERNARD on Feb. 2nd, 2010. Bought by Tsinghua University.

Robert Ayres (1932—) and Allen Kneese (1931—2001) presented the first version of what would become MFA of national economies as early as 1969. Ayres subsequently edited two books on resource accounting as well as *A Handbook of Industrial Ecology*. Understanding the structure and functioning of the industrial or societal metabolism is at the core of industrial ecology. MFA refers to the analysis of the throughput of process chains comprising extraction or harvesting, chemical conversion, manufacturing, consumption, recycling, and disposal of materials.

In 1991, Peter Baccini (1939—) and Paul H. Brunner (1946—) described a few of properties now common in contemporary MFAs in their discussion of anthropogenic flows of food, water, and other resources in a fictional region that they used for illustration. In 2004, Paul H. Brunner and Helmut Rechberger wrote the book entitled *Practical Handbook of Material Flow Analysis*, which has significantly enhanced the booming of theory and practices. It is the first-ever book on this subject, establishing a rigid, transparent and useful methodology for investigating the material metabolism of anthropogenic systems. In recent fifteen years, MFA methodology has been extended and improved, particularly with respect to dynamic MFA, uncertainty analysis & sensitive analysis, and supporting software. Thus, they presented the second edition entitled *Handbook of Material Flow Analysis: For Environmental, Resource, and Waste Engineers*.

Box 1-4: Paul H. Brunner (1946—)

Brunner is recognized as the father of material flow analysis. Together with Peter Baccini from ETH Zurich, he published the landmark book *Metabolism of the Anthroposphere*, presenting a new view of the interactions among human activities, resources, and the environment. He successfully promoted the application of material flow analysis for improving decision making in resource and waste management.

A great progress on industrial ecology in recent two decades was partly concentrated on resource and environmental sustainability. MFA is well used to support industry progress and environmental improvement. Some leading scientists in this area include Thomas E. Graedel, Braden R. Allenby, Yi Qian, and Zhongwu Lu. A great number of scholars are growing around the world.

In addition to national accounting of material flows, currently, MFA has been increasingly used as a basis for analyzing and planning the resource and environmental management. By search on the Web of Science Core ("material flow analysis" or "substance flow analysis" as abstract), more than 49 000 articles that used an MFA approach for analyzing the environmental and economic issues were found until the year of 2022.

Box 1-5: Thomas E. Graedel (1938—)

Graedel joined Yale University in 1997 after 27 years at AT&T Bell Laboratories and is currently an emeritus professor of Industrial Ecology at Yale. One of the founders of the field of industrial ecology, he coauthored the first textbook in that specialty. He was the inaugural President of the International Society for Industrial Ecology of 2002—2004 and winner of the 2007 ISIE Society Prize for excellence in industrial ecology research. He has served three terms on the United Nations International Resource Panel, and was elected to the U. S. National Academy of Engineering in 2002.

Publications in these areas started in the late 1990s and the number has gradually increased, to reach 3 983 in 2021. In addition to *Journal of Industrial Ecology*, plenty articles are found in journals such as *Environmental Science & Technology*, *Resources*, *Conservation and Recycling*, *Waste Management*, and *Waste Management & Research*.

MFA, originates from Society's Metabolism and is used in physical units (usually in the mass unit of *t*), and is to account materials from mining, production, conversion, consumption, recycling, to the final disposal. The objects for the analysis may include resources, energy, raw materials, products, wastes, and include single element. In 1970s and 1980s, the propose and improvement of Physical Balance, Industrial Metabolism and other theories laid a foundation for the study and practice of MFA in economic systems. As early as in 1966, W. Leontief used input-output model to describe the economic structure of the stock and flow. In 1969, R. U. Ayres first used "material balance principle" to examine the material flow of the national economy. On the basis of his study, I. Wernick and J. H. Ausbel advanced the basic National MFA framework of the United States. These are all early relevant exploration and study.

In early 1990s, the application of MFA of economic systems started in Austria, Japan and Germany, after which MFA became a rapidly developing science field. In 1990s, Wuppertal Institute raised Material Flow Accounts for the quantitative measurement of the amount of substance use in economic systems, and also proposed Ecological Rucksacks (ER), the concept called Hidden Flow (HF) later. In 1996, European Commission (EC) set up a coordinate accounts plan and its platform (www. conaccount. net) of the "concerted action". This is considered to be the first milestone of international cooperation on MFA. The second milestone is the comprehensive analysis of five national economic systems, including the United States, Japan, Austria, Germany, and the Netherlands. It was led by World Resource Institute (WRI) and carried out in two stages. In 1997, the first research paper was completed, which gave the result of the total material input and suggested relevant indicators to measure their status. In 2000, the second one was finished, in which the amount of total material output has been given, and relevant indicators for measure have been suggested, too.

In 2001, the European Union Statistical Office (EUROSTAT) published the first handbook on methodology of MFA for economic systems. From then on, the first international official guidelines, known as the "European Union (EU) guidelines", were widely applied. Since then, MFA was in considerable attention within Europe. EUROSTAT, European Environment Agency (EEA), EC and other institutions have been carrying out much of this work.

1.3 Biogeochemical cycle

1.3.1 Resource stock

The earth system (including the earth and its atmosphere) is an assemblage of atoms from 92 natural elements. Nearly all of these atoms have been present in the earth system since the formation of the earth 4.6 billion years ago by gravitational accretion of a cloud of gases and dust. The element exists but differs in spheres in the earth. The most abundant element in the universe is hydrogen, which makes up about three-quarters of all matter (Table 1 - 1). Helium makes up most of the remaining 25%. Oxygen is the third-most abundant element in the universe. All of the other elements are relatively rare. The chemical composition of earth is quite a bit different from that of the universe. The most abundant element in the earth's crust is oxygen, making up 46.6% of earth's mass. Silicon is the second most abundant element (27.7%), followed by aluminum (8.1%), iron (5.0%), calcium (3.6%), sodium (2.8%), potassium (2.6%), and magnesium (2.1%). These eight elements account for approximately 98.5% of the total mass of the earth's crust. Of course, the earth's crust is only the outer portion of the earth.

Table 1 - 1 Ten most abundant elements in earth's major reservoirs

Atmosphere		Hydrosphere		Biosphere		Lithosphere	
Element	Value	Element	Value	Element	Value	Element	Value
Nitrogen	780 840	Oxygen	857 000	Oxygen	523 400	Oxygen	464 000
Oxygen	209 500	Hydrogen	108 000	Carbon	302 800	Silicon	282 000
Argon	9 300	Chlorine	19 000	Hydrogen	67 500	Aluminum	82 300
Carbon	90.05	Sodium	10 500	Nitrogen	5 000	Iron	56 000
Neon	18.18	Magnesium	1 350	Calcium	3 700	Calcium	41 000
Krypton	1.14	Sulfur	885	Potassium	2 300	Sodium	24 000
Helium	5.24	Calcium	400	Silicon	1 200	Magnesium	23 000
Hydrogen	0.55	Potassium	380	Magnesium	980	Potassium	21 000
Xenon	0.09	Bromine	65	Sulfur	710	Titanium	5 700
Sulfur	0.05	Carbon	28	Aluminum	555	Hydrogen	1 400

Note: The composition given for the biosphere includes water in living organism; the composition for the lithosphere is that of the crust. In addition to the single-element gases tabulated here, the atmosphere also contains 0.25% H_2O, 1.79 ppmv CH_4, 0.325 ppmv NO, 400 ppmv CO_2, 0.002% SO_2, 0.02 ppmv NO_2, 0.000 06% H_2S, and 0.04ppmv O_3. All abundances are in parts per million (ppm).

Natural biogeochemistry and social-economy metabolism, such as trade and logistics, make up material flow in the macroworld. From global perspective, the spatial landscape of typical metals has changed dramatically since 1960. As anthropogenic stock, geological metals minerals, for example, have been dramatically removed from subsurface to aboveground (i. e. in-use stock and waste). Currently, the twelve most common metals, such as Ag, Au, Bi, Cd, Cu, Fe, Hg, In, Pd, Sb, Sn, and Zn, have a visible stock on the surface, indicating that the bulk of these metals have been extracted from the geological reserve (i. e. economically extractable resources, Figure 1 - 6). In-use stocks for a few metals have been estimated regionally and globally. The enrichment of metals in modern society and the importance of the realization of existing reservoirs created by human activities.

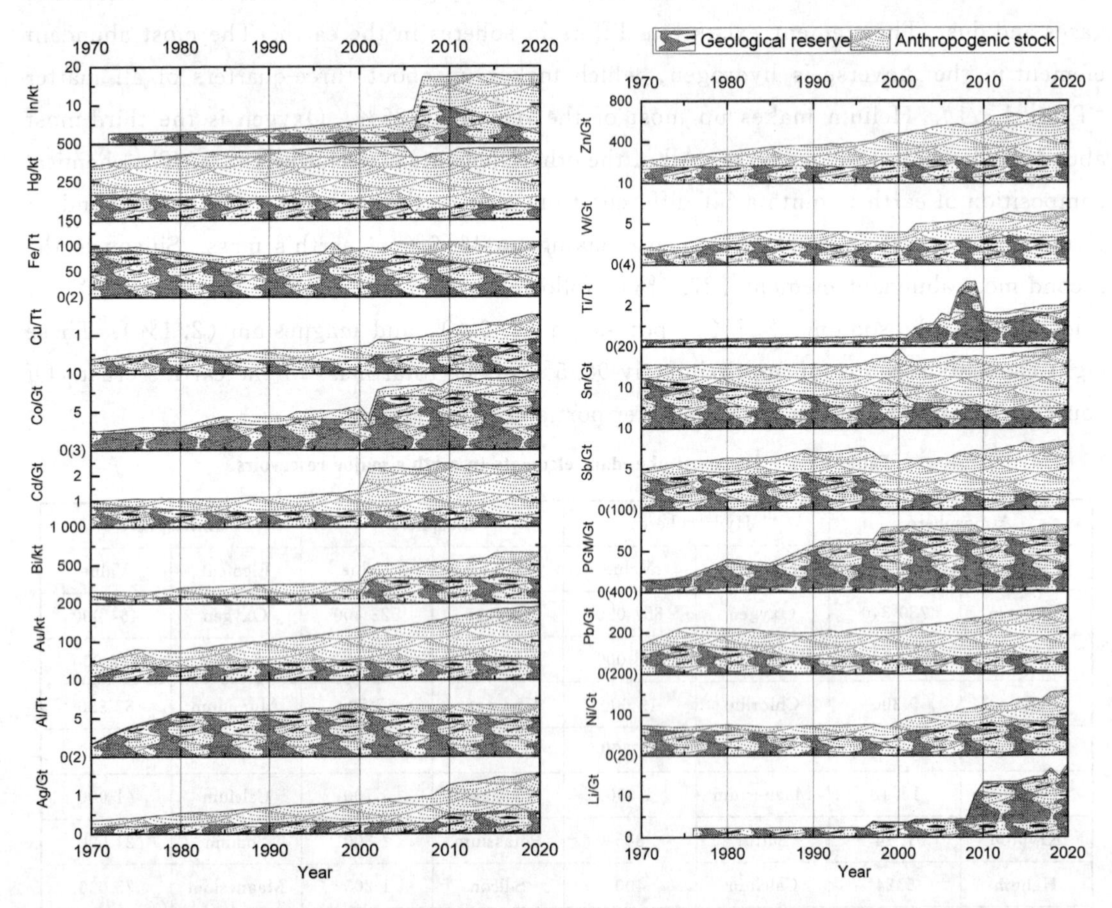

Figure 1 - 6 The global evolution of spatial landscape of typical metals in 1970—2020[①]

While anthropogenic stock of these metals in 2020 accounted for over 70% of total geological and anthropogenic resource, some geological metals like In (88%), Cd (87%), Hg (84%), Sb (82%), Sn (76%), Au (74%), and Pb (73%), are rapidly depleting. But China demonstrated a more severe situation than global circumstance, in particular on Co, Pb, Ni, and Al. Furthermore, trade plays an important function in balancing the regional landscape. China, for example, is the global manufacturing hub because it imports resources

① Data of geological reserve source from USGS. Anthropogenic stock, consisting of in-use, waste, and even landfill, is the sum of previous production. 1 Tt= 10^3 Gt= 10^6 kt $=10^9$ t.

from Australia and exports the finished product to the United States. Individual mining, manufacturing, and recycling companies might pursue resource utilization and waste minimization in the microworld.

In addition, a total of the rare earths element (REE) totaled around 440 kt in 2007, with most of the stock in four elements: La, Ce, Nd, and Pr. That stock is some four times the 2007 annual extraction rate, which suggests that REE recycling may have the potential to offset a significant part of REE virgin extraction in the future. Valuable materials are contained as stock in various electronics (Table 1 - 2). In particular, precious metals like silver, gold, and palladium are much more abundant in electronics than in the lithosphere.

Table 1 - 2 Weight distribution of different materials in various electronics

Weight	Fe/%	Al/%	Cu/%	Plastics/%	Ag/ppm	Au/ppm	Pd/ppm
CRT-board	30	15	10	28	280	20	10
Mainframe-board	7	5	18	23	900	200	80
Mobile phone	7	3	14	43	3 000	320	120
Portable audio	23	1	21	47	150	10	4
DVD-player	62	2	5	24	115	15	4
Calculator	4	5	3	61	260	50	5

1.3.2 Cycling of elements

Biogeochemical cycles mainly refer to the movement of nutrients and other elements between biotic and abiotic factors. The matter on earth is conserved and present in the form of atoms. Since matter can neither be created nor destroyed, it is recycled in various forms in the earth system.

It is one of the biogeochemical cycles in which carbon is exchanged among the biosphere, geosphere, hydrosphere, atmosphere and pedosphere. All green plants use carbon dioxide and sunlight for photosynthesis. As a result, carbon is thus stored in the plant. The green plants, when dead, are buried into the soil that gets converted into fossil fuels made from carbon. When burned, these fossil fuels release carbon dioxide into the atmosphere (Figure 1 - 7).

At the same time, plant-eating animals obtain the carbon stored in the plants. When the animals die and decay, the carbon returns to the atmosphere. Carbon is also returned to the environment through cellular respiration in animals. Much of the resulting carbon, in the form of carbon dioxide, is stored in the form of fossil fuel (coal and oil) and can be extracted for a variety of commercial and non-commercial purposes. When factories use these fuels, the carbon is again released back in the atmosphere during combustion.

Extending to the society, there are three subsystems: the economic or technological

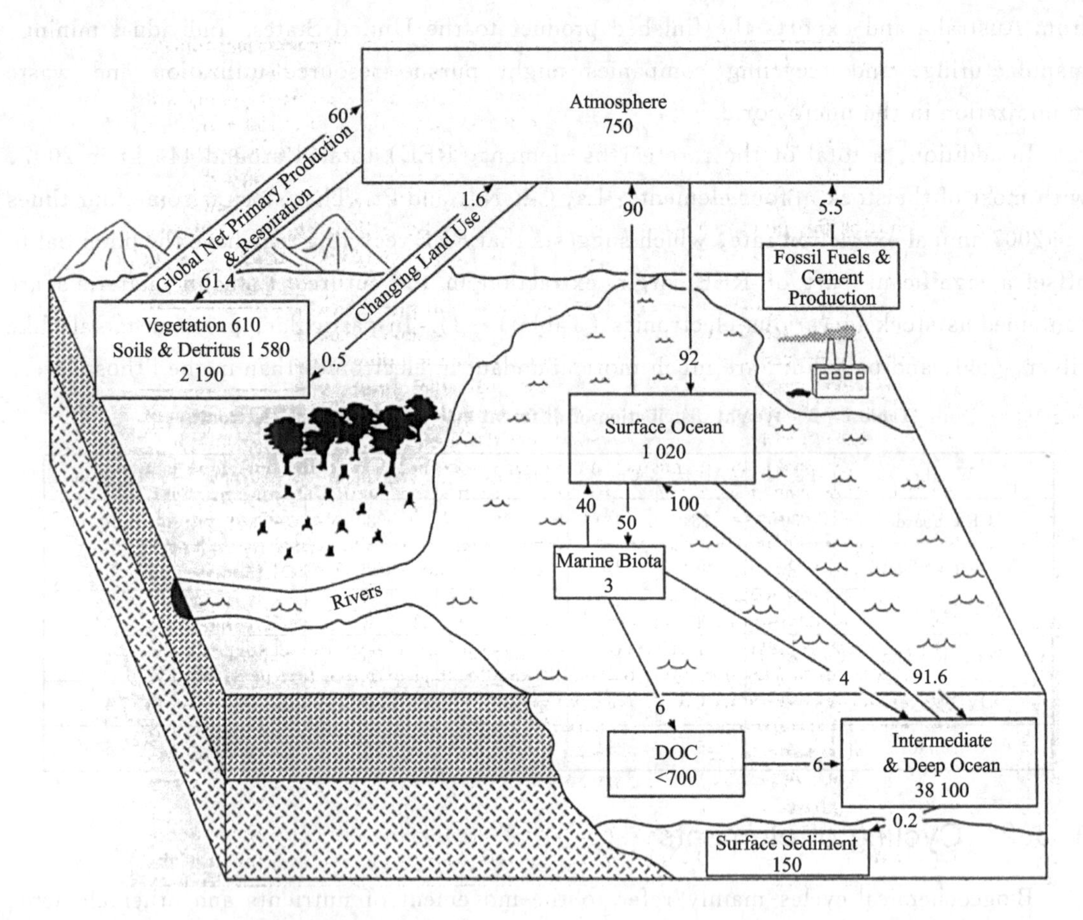

Figure 1 - 7 The global carbon cycle showing the stock reservoirs (in Gt C) and fluxes (Gt C/y) relevant to the anthropogenic perturbation as annual averages over the period 1980—1989①

domain, the environment or biosphere, and the lithosphere, i. e. the ultimate source and sink of the elements like metals (Figure 1 - 8, stocks of substances are depicted as bold squares, flows as arrows, environmental media as parallelograms and all stages in the economy as squares). Stocks in the lithosphere are assumed to be immobile, while all stocks in the economy and environment are mobile. To analyze the structure of the economy the processes have been grouped into five stages: extraction and refining, covering both functional and non-functional mining and refining of metal, as well as secondary recovery of metal scrap; production, which covers the production of both functional and non-functional applications of metals (e. g. copper wire and cadmium in fertilizer, respectively); use, which includes, for example, storing soft drinks in aluminum cans and applying fertilizer in agriculture; waste management, comprising collection, sorting, treatment and storage of metals in municipal and industrial waste.

① Source from Schimel (1995).

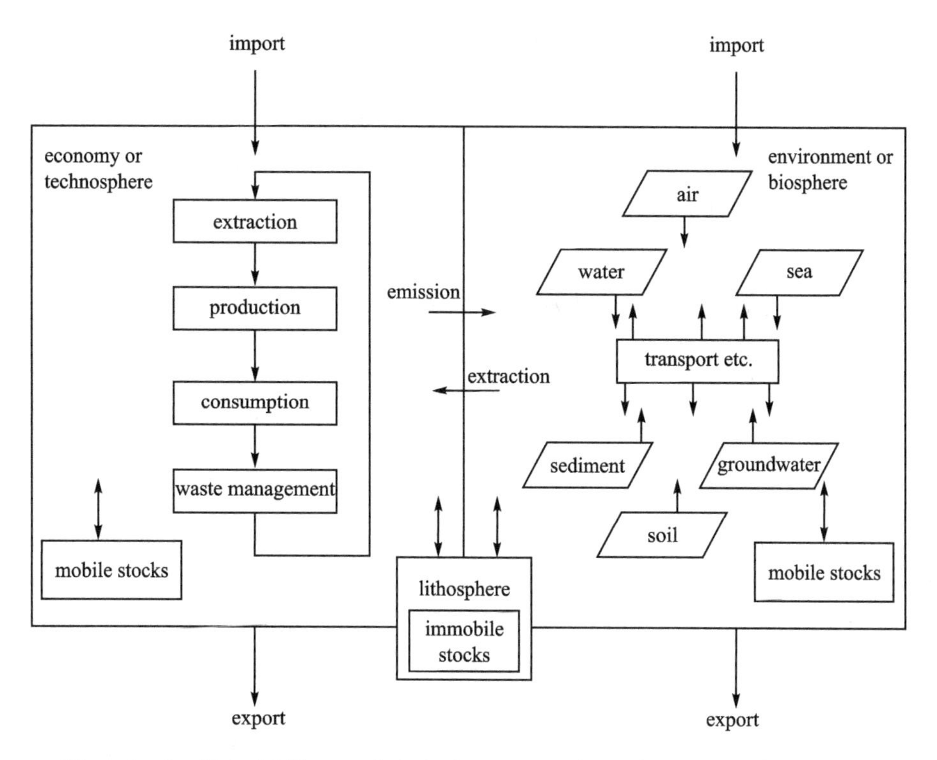

Figure 1 - 8 Material flows and stocks in economy, the environment and lithosphere

1.4 Industrial ecology and anthropogenic metabolism

1.4.1 Industrial ecology

By analogy with natural ecology, industrial ecology is the study of material and energy flows through industrial systems. The global industrial economy can be modelled as a network of industrial processes that extract resources from the earth and transform those resources into consumable energy products which can be bought and sold to meet the needs of humanity. Industrial ecology seeks to quantify the material flows and document the industrial processes that make modern society function. Industrial ecology is therefore a study of technological biology, its use of resources, its potential impact on the environment and how to rebuild its interaction with the natural world for global sustainability.

Industrial ecologists are often concerned with the impacts that industrial activities have on the environment, with the use of the planet's supply of natural resources, and with problems of waste disposal. Industrial ecology is an emerging discipline that began to emerge in 1990s as an emerging multidisciplinary research area combining engineering, economics, sociology, toxicology and natural sciences.

In natural ecological science, the food chain is a network of linear links in the food chain, starting with the production of organisms (such as grass or trees, which use solar radiation to produce food through photosynthesis) and ending with apex predator (such as grizzly bears or orcas), insectivores (such as earthworms or lice) or decomposer species (such as fungi or bacteria). In industrial ecosystems, as in food chain in natural ecosystems, physical flows occur from miners (geological mining), manufacturers, producers, distributors, consumers, collectors, and recyclers (e. g. decomposers). These analogies allow the classical laws and principles of natural ecology to be transplanted into industrial ecology.

1.4.2 Anthropogenic metabolism

Physiologically, metabolism is the chemical (metabolic) process that occurs when the body converts food and drink into energy. Metabolism, like the analogy between natural and industrial ecology, can be used at both the urban and social levels. Urban metabolism, proposed in 1965, has become an effective method to evaluate the flows of energy and materials within an urban system, thereby providing insights into the system's sustainability and the severity of urban problems such as excessive social, community, and household metabolism at scales ranging from global to local. The definition of urban metabolism has mostly emphasized the technological metabolic processes of a socioeconomic system and has treated natural systems as nothing more than a kind of fuel or support for these processes, instead of combining natural or organic processes into urban metabolic processes or isolating natural systems as distinct components that also require study. Urban metabolism can therefore be considered as one part of industrial ecology and as a metaphor for the application of urban systems.

In the anthroposphere, the current human activities have been expanded from the household to the world (Figure 1 - 9). The notion metabolism is used to comprehend all physical flows and stocks of matter and energy within the anthroposphere. At the different scales, household metabolism, neighborhood metabolism, urban metabolism, regional metabolism, social metabolism, and anthropogenic metabolism may exist to indicate the performance of metabolic flow (Figure 1 - 10). In this context, metabolic flows refer to the flows of materials, water, energy, and nutrients and include both "resource" flows and "waste" flows. "Resources" refer to materials, energy, information, and services that can be used to produce some tangible (e. g. materials) or intangible (e. g. information) product. "Wastes" refer to the byproducts of a component of the system that are not reused by that component. These "wastes" often become important resources both for the component itself (e. g. recovery of waste paper by pulp and paper plants) and for other components (e. g. burning paper waste by another industry to generate energy). Ultimately, anthropogenic metabolism is an important educational and decision-making tool that, as data availability increases, can provide important information about human activities and their sustainability goals.

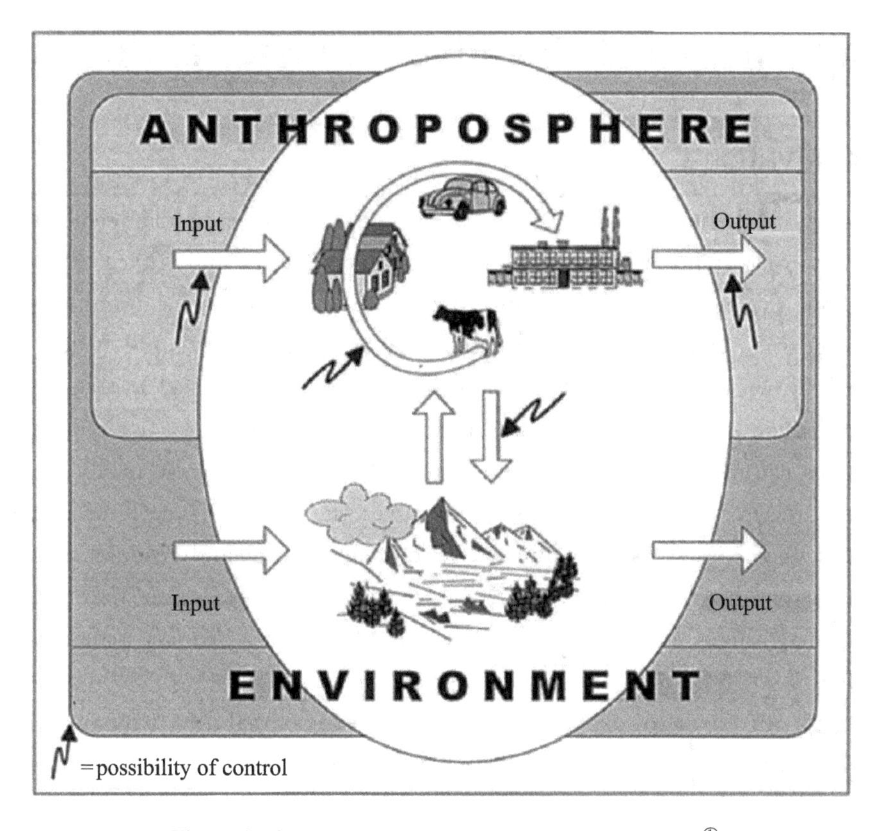

Figure 1 - 9 The anthroposphere and the environment①

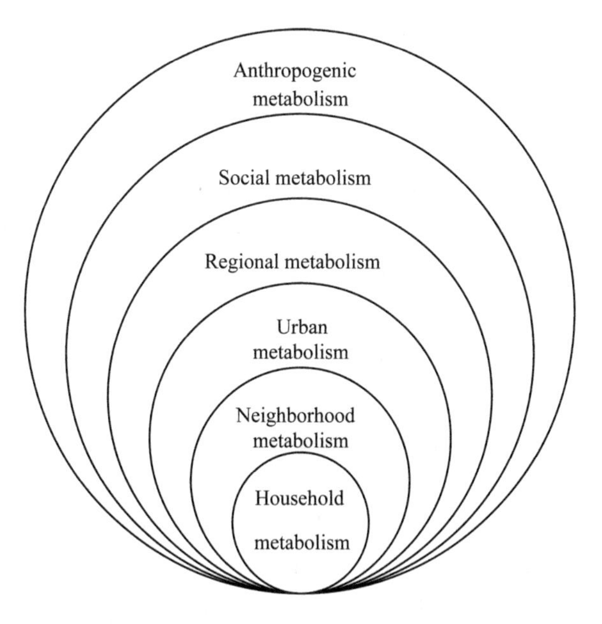

Figure 1 - 10 The scales of anthropogenic metabolism

① Source from the reference [21].

1.5 Motivation and objective for material flow analysis

1.5.1 Motivation

MFA is one of the most widely accepted and used tools in industrial ecology discipline, measuring input-output materials and examining the pathways and fluxes of each material flow throughout the system. At the same time, MFA can track the number of materials entering the economy and environment at all stages of a commodity's life cycle.

It is well known that ecological environment problems are caused by over-exploitation, extensive utilization and over-consumption of resources. How to investigate ecological problems? How can resource sustainability issues be identified? These problems are largely related to resource flows, which in most cases cover the entire life cycle of materials. For example, how does copper affect China's environmental risks and resource sustainability? Copper is not only a heavy metal susceptible to environmental pollution, but also a precious metal used in many high-tech industries. The exploitation of its primary geological reserves, whether domestic or imported, can meet the domestic demand. It will be supplied to China and other countries through exportations. Resource sustainability depends on domestic reserves and international importations. How much does the flow of mining support the demand for industry development? Copper, on the other hand, can be released into the environment during mining, manufacturing and even recycling. How to control the process to reduce releases?

All of these questions inspire us to think about the material flow: the status and evolution of each program. Copper importations, for example, play an important role in China's copper supply. Quantities and future trends in importations from some of the countries or regions concerned can be answered by means of material flow diagram. A full MFA map has been prepared, which clarifies some of the potential problems and outlines possible solutions. Thus, the motivation for seeking the adoption of multi-framework agreements is naturally this way.

1.5.2 Objective

MFA analysis is a targeted, responsible and efficient analysis of physical and substance flows within the socio-industrial production systems. Logistic management aims to conserve natural resources and create sustainable production. A distinction is made between material flow management at the national or regional level, which focuses more on the creation of ecologically sustainable cycles aimed at the protection of the environment and efficient use of resources. In industrial or operational material flow management, consideration of energy and material flows should contribute to the optimization within production systems.

Within the framework of logistics management, the tools of the Ministry of MFA are also used. The goal of MFA is to map, analyze, and evaluate all streams in the definition system. The focus is on physical (or energy) flows and their impact on the environment. The origin of the MFA is to be found in waste management, since it became necessary to close cycles and reduce the amount of waste.

As far as management policies are concerned, MFA can be used to identify, prioritize, analyze, and improve the effectiveness of measures at an early stage, as well as to design effective material management strategies based on sustainability. MFA have high potential as guidance tools at the regional level, for example as part of regional environmental management and audit system or as part of Local Agenda 21 processes. Material management based on MFA complements traditional environmental and resource management strategies, which tend to focus on specific environmental components and measure concentrations of substances in various media. Instead, the MFA provides an overview of the entire system by linking the human sphere, part of the biosphere in which humans' activities are operated to the environment. This systematic approach shifts the focus from so-called "filtering strategies" at the back end to more proactive front-end measures. The Foreign Office reviews short-term and long-term loadings, rather than concentrations, and highlights current and potential material accumulations, known as material stocks. These represent either potential environmental problems (e. g. large stocks of hazardous materials) or potential sources of future resources (e. g. urban mining, i. e. recycling from waste).

In this way, MFA can assist precautionary policy making by highlighting future environmental or resource issue problems without relying on signals of environmental stress. The objective of materials management is: firstly, to analyze material flows and stocks; secondly, to evaluate these results; and thirdly, to control material flows in view of certain goals such as sustainable development. MFA is an excellent tool for the first objective and is well suited to generate a base for the other two objectives. MFA results can be compared against environmental standards or can be interpreted using assessment or indicator methodologies (e. g. environmental impact assessment or ecological footprint). Finally, MFA is objective to trace the flow of raw materials through the system, retrace waste to the point where it is generated, polish data in a decision-oriented way, identify weaknesses in the production process, and set priorities for appropriate measures aimed at minimizing waste and emissions.

1.6 Further reading

[1] Brunner P, Rechberger H. Practical Handbook of Material Flow Analysis[M]. Boca Raton: Lewis Publishers and CRC Press LLC, 2004.

[2] Brunner P H, Rechberger H. Handbook of Material Flow Analysis: For Environmental, Resource, and Waste Engineers[M]. 2nd ed. Boca Raton: CRC Press, 2016.

[3] Baccini P, Brunner P H. Metabolism of the Anthroposphere[M]. [S. l.] Springer-Verlag, 1991.

[4] Ayres R U, Ayres L. A Handbook of Industrial Ecology[M]. Cheltenham: Edward Elgar Publishing, 2002.

[5] Baccini P, Brunner P H. Metabolism of the Anthroposphere: Analysis, Evaluation, Design[M]. Cambridge: MIT Press, 2012.

[6] Graedel T E. The evolution of industrial ecology [J]. Environmental Science & Technology, 2000, 34(1):28-31.

[7] Collier P, Alles C M. Materials ecology: an industrial perspective[J]. Science, 2010 (330):919-920.

[8] Kral U, Lin C Y, Kellner K, et al. The copper balance of cities [J]. Journal of Industrial Ecology, 2014(18): 432-444.

[9] Sverdrup H, Ragnarsdóttir K V. Natural resources in a planetary perspective [J]. Geochemical Perspectives, 2014, 3(2):129-341.

[10] Weisz H, Suh S, Graedel T E. Industrial ecology: the role of manufactured capital in sustainability[J]. Proceedings of the National Academy of Sciences, 2015 (112): 6260-6264.

[11] Graedel T E. Material flow analysis from origin to evolution [J]. Environmental Science & Technology, 2019(53): 12188-12196.

1.7 Exercises

1. What is the MFA concept? What is the difference between MFA and SFA?

2. How can the scope of material flow analysis be understood on a temporal and spatial scale?

3. What is the industrial ecology? What is the future of the technology-environment relationship?

4. What are the scales of anthropogenic metabolism? Please read more publications to illustrate the different scales.

5. Cobalt is a metal widely used in the electronic industry and batteries. Evaluate cobalt's availability from the standpoint of abundance, co-occurrent, and geographic occurrence. Based on whatever information, you can locate on cobalt, what do you predict for lead as an industrial material in the next few decades?

6. The grade of copper ore has decreased over time. Assuming a 12% loss of copper to tailings in the mill and a 1% loss of copper to slag in subsequent smelting. If you wished to produce 1 kg of copper, how much ore would be required in 1990 and 2020, respectively, if the loss rate did not change?

7. More than $ 100 million was invested in one waste incineration plant, and still, the objective of the treatment, namely, to produce recycling materials of a certain quality, could not be reached. Engineers, plant operators, waste management experts, financiers, and representatives from government want to find the means to achieve the goal. Could you devise an approach to solve the problem?

8. Read the latest publications of *Environmental Science & Technology*, *Resources, Conservation and Recycling*, or *Journal of Industrial Ecology*, and look for recent applications of MFA. Identify the goals of the MFA, the procedures, and the results, and discuss whether the conclusions and implications have been appropriately visualized by MFA. Evaluate whether MFA was the only possible approach to reach the objectives or whether other methods would have allowed the same conclusions.

9. China started the zero-waste city construction around the China in 2019. Please use urban metabolism to uncover the process how to achieve the zero-waste city. What are the potential problems in this process?

Chapter 2 Methodology of Material Flow Analysis

How to achieve a complete MFA diagram is of vital importance to illustrate one systematic resource or environmental problem. The methodology has been growing from lots of practices and studies. Some terms and procedures have been well designed by Prof. Paul H. Brunner and Prof. Helmut Rechberger. MFA is an increasing tool and philosophy while more and more persons were involved from government to academic. Some basic theory, validating method, and new software have been created in recent years. This chapter will address the basic term and theory of fundamental science, MFA procedure, uncertainty & sensitivity analysis, and main software. More time could be needed for fresh users to deeply master this methodology.

2.1 Basic term and theory of fundamental science

2.1.1 Basic term and definition

Material flow analysis (MFA) is a systematic assessment of the flows and stocks of materials within a system defined in space and time. MFA involves a lot of terms mainly like substance, material, flow, flux, process, stock, and boundary.

1. System, substance, and material

System isa regularly interacting or interdependent group of items forming a unified whole, and here it is defined as the whole research object. Substance is regulated as the pure matt in chemical science. It consists of the elementary and chemical substance, for instance, copper (Cu) and carbon dioxide (CO_2), respectively. Material is a physical substance that things can be made from. For instance, plastic and waste are the typical material. Iron, aluminum, and copper are the most frequently investigated substance or material. The majority works consider metal use in some or all of the following end-use sector categories: transportation, buildings and construction, infrastructure and telecommunication, machinery, electric appliances, consumer goods, containers and packaging.

2. Flow and flux

Flow is the movement of something in one direction, covering inflow and outflow (Figure 2-1). In the traditional knowledge, flow is well employed in the hydrodynamics. In MFA, flow can be expressed, for example, as the flow of traffic and product importation. Flow is the volume of material passing through a cross section per unit of time. Flux is the volume of material passing through one-unit cross section per unit of time. A flux is often defined as a flow per "cross section". In MFA, commonly used cross sections are a person, 1 m^2 of a surface area, or an entity such as a private household or enterprise. The flux might be given in units like $kg/(s \cdot m^2)$. For instance, how many mobile phones demand in Beijing in the year of 2020? Here the system and boundary are Beijing, the time is 2020, the mobile phones demand will be the flow, and the mobile phones demand per capita will be the flux while the cross section is the population (Figure 2 - 2).

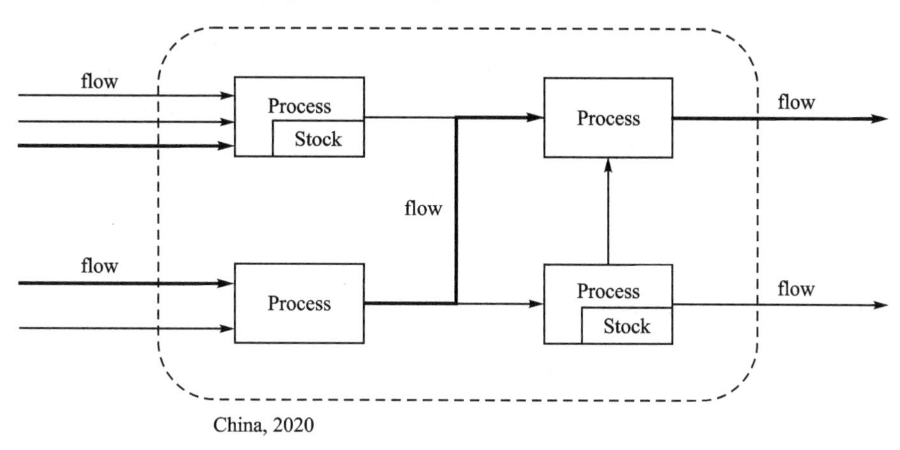

Figure 2 - 1 Typical symbols used in MFA diagram

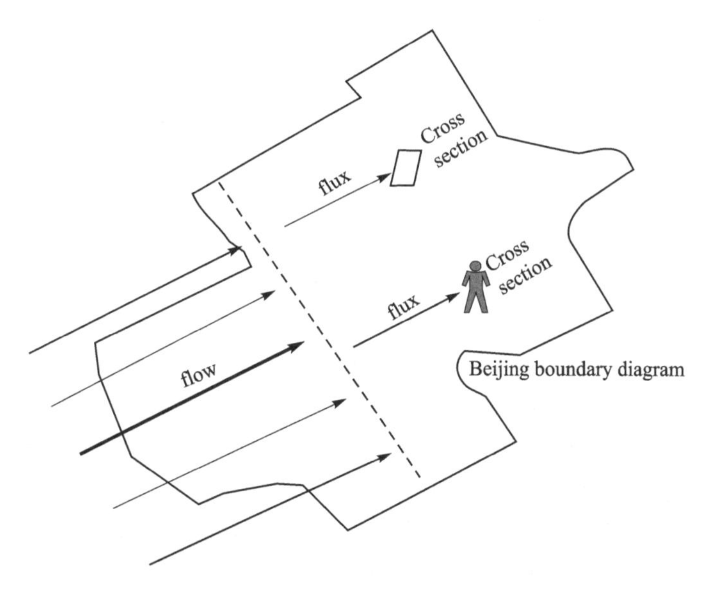

Figure 2 - 2 The illustration of flow and flux with mobile phones demand in Beijing

3. Process, stock, and boundary

Process isa series of action that happen, which links the inflows and outflows (Figure 2 - 1). For instance, generation and recycling are process. MFA of metals cover the whole life cycle of a metal from primary mining to raw material production to product manufacturing to use and finally waste management. Stock is a sink or stored amount of material in one phase. For instance, copper reserve is the stock of copper in the lithosphere. Boundary can be classified as two types of temporal and spatial scale. Again, how many mobile phones demand in Beijing in the year of 2020? Beijing is the spatial boundary, and 2020 is temporal boundary. The spatial extent ranges from urban to global system boundaries.

In light of the temporal boundary, MFA can be classified as static MFA and dynamic MFA. A static MFA is a single time (e.g. year, week) evaluation of product and material flows with a fixed product lifespan, whereas a dynamic MFA is a multi-year evaluation of product and material flows with variable product lifespans. In another words, an MFA is static if it describes a "snapshot" of a system in time. An MFA is dynamic if it describes the behavior of a system over a time interval. For instance, how about the generation, consumption, and waste of mobile phones in Beijing in the year of 2020 belongs to the static MFA, and how about the generation, consumption, and waste of mobile phones in Beijing during 2010—2020 belongs to the dynamic MFA.

2.1.2 The law of conservation of mass

The law of conservation of mass is the principle that states that neither physical transformation nor chemical reactions create or destroy mass in an isolated system. According to this principle, the reactants and products in a chemical reaction must have equal masses. Therefore, the sum of masses of wax and oxygen (reactants) in a chemical reaction must be equal to the amount of the masses of carbon (Ⅳ) oxide and water (products). The law of conservation of mass is essential in calculations involving the determination of unknown masses of reactants and products in any given chemical reaction.

The law of conservation of mass originated from the proposal by an ancient Greek that the total amount of matter in the universe does not change. In 1789, Antoine Lavoisier termed the law of conservation of mass as the vital principle in physics. Einstein later amended this law by including energy in its description. According to Einstein, the law became the law of conservation of mass-energy, which states that the total mass and energy does not change in any given system. From this principle, energy and mass can be converted from one to the other. Nevertheless, because energy consumption or production in common chemical reactions accounts for a negligible amount of mass, the law of conservation of mass is still a fundamental concept in chemistry.

An MFA connects the sources, the pathways, and the intermediate and final sinks of a material. Because of the law of the conservation of matter, the results of an MFA can be controlled by a simple material balance comparing all inputs, stocks, and outputs of a process. It is this distinct characteristic of MFA that makes the method attractive as a decision-support tool in resource management, waste management, and environmental management.

An MFA delivers a complete and consistent set of information about all flows and stocks of a particular material within a system. Through balancing inputs and outputs, the flows of wastes and environmental loadings become visible, and their sources can be identified. The depletion or accumulation of material stocks is identified early enough either to take countermeasures or to promote further buildup and future utilization. Moreover, minor changes that are too small to be measured in short time scales but that could slowly lead to long-term damage also become evident.

2.1.3 The law of anthropogenic circularity

The development of many-scale disciplines in terms of green chemistry, closed-loop supply chain theory, and industrial ecology are boosting the emergence and evolution of anthropogenic circularity science. Anthropogenic circularity is defined as the combination of biogeochemical cycle and anthropogenic recycling, which is the fundamental rule to uncover the circular economy for the sustainable development (Figure 1 - 2). Three principles are expressed as first, second, and third laws of anthropogenic circularity from the macroscopic and mesoscopic scales to the microscopic scale.

1. First law of anthropogenic circularity

The first law of anthropogenic circularity (Eq. (2 - 1)) also known as the fundamentals of constant abundance, states that the relative contents (or abundance) of various moved chemical elements in the universe are constant.

$$F = \frac{Ma}{\sum Mn} \tag{2-1}$$

where F is the abundance of elements (%), Ma is the content of element (tons), Mn is the total sum of all the elements' content in tons.

One of the persistent problems, which interest to environmental and allied scientists concerned with the chemistry of meteorites and planets, has been the original composition of the solar system. When the abundance of elements is constant, the total mass of all elements in the universe will be constant.

The first law of anthropogenic circularity indicates the importance and scope of anthropogenic circularity science. The earth's natural resources are limited and material supply from urban mining is much needed to meet the shortage of virgin resources. The scope of anthropogenic circularity covers the 83 types of natural elements. With anthropogenic circularity, any waste containing any of the 83 elements is considered as a raw material to recycle for new products.

2. Second law of anthropogenic circularity

The second law of anthropogenic circularity is the principle of material cycling, which is a universal law of ecological systems. It means that all the elements and products made from them are moving along cyclical routes in an ecological system.

(1) Composition and structure of the ecosystem

Driven by the biological, physical and chemical components, the ecosystem consists of two main components, i. e. community, which is called biocoenosis and habitat (biotope). Functionally, the two components of the ecosystem (autotrophs and heterotrophs) can be recognized usually with four constituents, i. e. abiotic, producers, consumers and decomposers. The four constituents play different roles. For example, producers (e. g. plants and algae) acquire nutrients from inorganic sources which are supplied primarily by decomposers whereas decomposers, mostly fungi and bacteria, acquire some elements from organic sources that are supplied primarily by producers.

(2) Material cycle

According to the scope, the biogeochemical cycles consist of a geochemical and biological cycle. The geochemical cycle is the pathway of compounds and elements, covering the adsorption in biont (a discrete living organism that has a specified mode of living), back to the environment via dying, residue, or excrement of biont, and further utilization for biont through five spheres, i. e. atmosphere, hydrosphere, pedosphere, lithosphere, and biosphere. The geochemical cycle supplies the external environment for the biological cycle. Biorecycling is the way to capture elements in ecosystems through adsorption, further utilization, decomposition, and new adsorption.

Material cycling is characterized by its indestructibility and circularity. During transformation, material and energy can be changed in form, but can never be destroyed. But the material cycle differs from energy flow because energy transfer is irreversibly leaving the ecosystem, whereas material can be recycled. Material is limited and heterogeneous in distribution can be recycled frequently owing to its perpetual utilization in the ecosystem.

The second law of anthropogenic circularity uncovers the direction from the environment (e. g. atmosphere), producer (e. g. greenery), consumer (e. g. animal, human), and decomposer (e. g. microbes). Therefore, this law is also known as the law of conservation of mass. Material can be changed in form but can never vanish during the cycle. The second law of anthropogenic circularity illustrates the feasibility of recycling.

3. Third law of anthropogenic circularity

The third law of anthropogenic circularity (Eq. (2 - 2)) is the principle of zero waste that chemical reactions should be designed to tentatively achieve zero emission of waste.

$$\sum W \to 0 \tag{2-2}$$

where W is the amount of waste.

The third law can also be explained via the principle of atom economy or atom efficiency, which is the conversion efficiency of a chemical process in terms of all atoms involved and the desired products produced (Eq. (2 - 3)). Atom economy is an important concept of green chemistry, and a widely used metric for measuring the "greenness" of a process or synthesis. However, it must be noted that not all 100% atom efficient processes can automatically be considered to be green.

$$E = \frac{D}{T} \times 100\% \tag{2-3}$$

Ideally, the manufacturing process would design some related reactions that would involve all atoms in the structure of the product. Through mass conservation, the total molecular weight of the reactants is equal to the total molecular weight of the product. In an ideal chemical process, the number of starting materials or reactants is equal to the amount of all the products needed and no atoms are wasted.

The third law of anthropogenic circularity at the microscopic level implies the importance of maintaining ecological balance. Natural material recycling is a process in which waste emission are virtually zero. If this process is disrupted and/or ecological balance is disrupted, waste will inevitably be generated, potentially endangering human life or environmental health.

2.2 Procedure of material flow analysis

MFA is growing and forming the procedure. The detailed methodological steps of MFA are described in references [27] and [28]. The approach for a material flow analysis is relatively simple and can be conducted in the following four steps (Figure 2 - 3): ① Identifying the problem with specification of the objectives and system determination; ② Creating the framework from process selecting, framework defining, and data collecting & inventory; ③ Estimating the unknown data; ④ Mapping MFA diagram and illustrating the results.

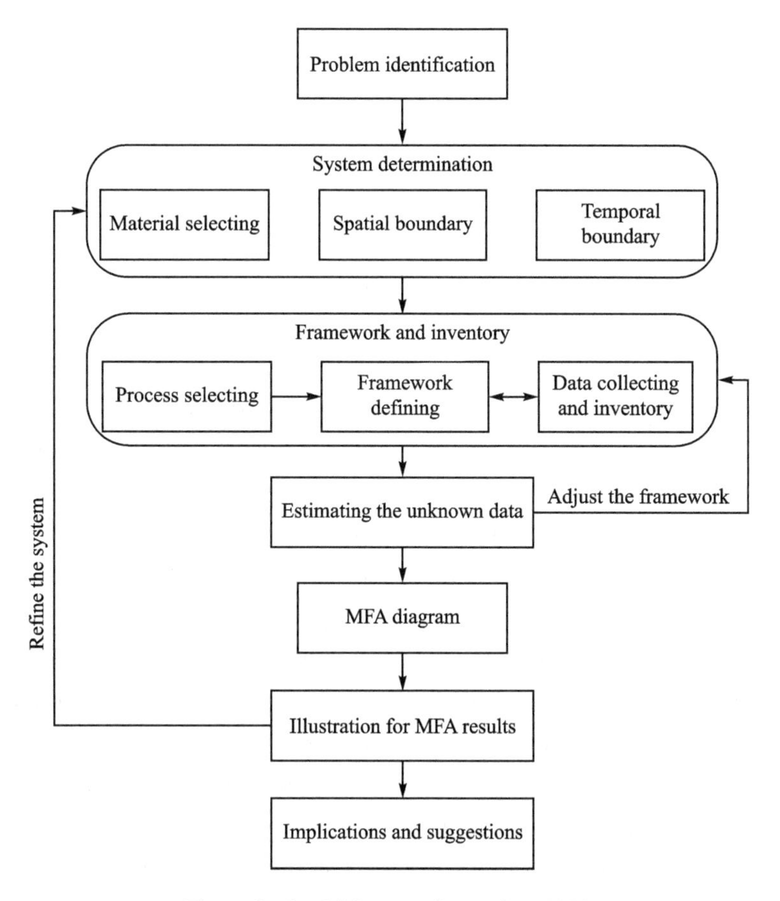

Figure 2 - 3 Main procedures of an MFA

2.2.1 The objectives and system determination

"Objectives and system determination", where the objective of the study is formulated; the scope is defined in terms of temporal, spatial, and technological coverage; and the level of sophistication in relation to the objectives is fixed. Additionally, the product(s) of study are described and the functional unit is determined.

The system to be investigated is defined by boundaries in space and time. The relevant processes, goods and substances are defined and linked. The selection of the processes involves identifying those key processes which most efficiently represent and describe the complex system under investigation. This is one of the most critical and demanding steps of an MFA, as one is trying to depict reality in a simplified manner. The selection of goods and substances is dependent on the nature of the study. In research-orientated projects, where the interest may be on understanding the urban metabolism of a region, indicator materials such as carbon and nitrogen (essential elements for the biosphere) and the elements lead, iron, aluminum and zinc (some of the most critical metals of the anthroposphere) are typically selected. In more applied investigations, where the interest may be on a specific environmental issue, problem-specific materials are usually selected. For example, forestry management issues may focus on timber, while eutrophication issues may focus on nitrogen and phosphorus as indicator materials.

2.2.2 The framework and data inventory

This section describes the MFA framework used to map the product flows through collection and subsequent EoL pathways within a given facility (Figure 2 - 4). The dashed line represents the system boundary of the specific study. The "inventory analysis", which results in a table that lists inputs from and outputs to the environment ("environmental interventions") associated with the functional unit. This requires the setting of system boundaries, selection of processes, collection of data, and performing allocation steps for multifunctional processes (e. g. a power plant producing energy not only for a single product).

In general, the complicated MFA can be divided into a few of simplified units. Each unit has some materials or goods input, a little stock, and additional output via the process (Figure 2 - 5). The collected data means to measure or estimate properties of an entity. There are several methods available that can be divided into direct and indirect methods. Typically, the direct methods cover: ① collect the sampling for measurement likely in a chemical way; ② collect the data from publications (e. g. statistics, article, website, margarine, and report); ③ interview some experienced experts to know some information used frequently in social science. If direct measurements are not available, indirect method could be employed to fill data gap in an MFA. For instance, if we want to know China's imported amount of plastic in 2000, but no data can be found in this year. In some cases, we can use the data of 1999 or 2001 (if available) to replace the data of 2000.

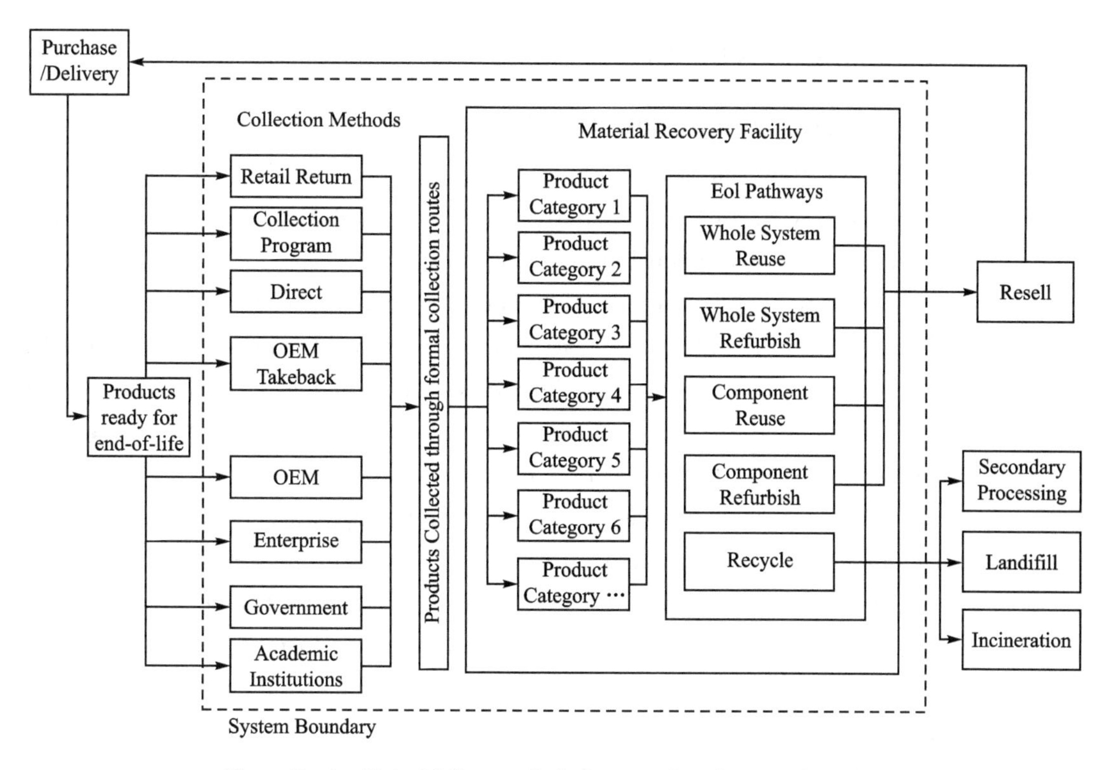

Figure 2 - 4 Material flow analysis framework and system boundary

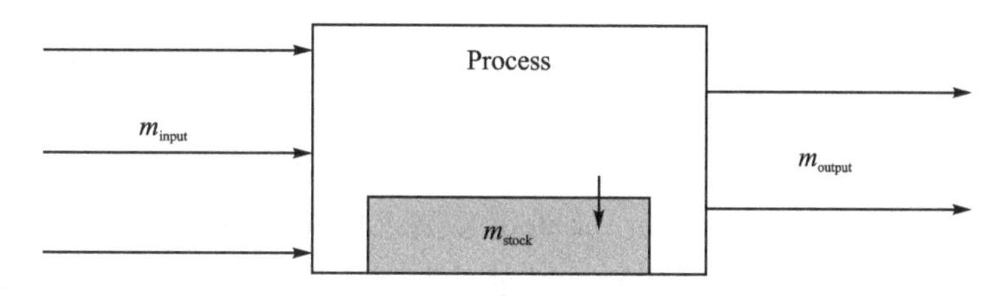

Figure 2 - 5 A simplified unit of MFA

The data inventory or revised inventory should cover all the considered input, stock, and output of all the flows in an MFA system. Finally, the data inventory for each flow is determined in Table 2 - 1.

Table 2 - 1 Data inventory for each flow in an MFA

Materials	Flow rate/($t \cdot y^{-1}$)	Concentration of substance/%				Substance flow rate/($kg \cdot y^{-1}$)			
M_1	m_1	c_{11}	c_{12}	…	c_{1n}	X_{11}	X_{12}	…	X_{1n}
M_2	m_2	c_{21}	c_{22}	…	c_{2n}	X_{21}	X_{22}	…	X_{2n}
…	…	…	…	…	…	…	…	…	…
M_k	m_k	c_{k1}	c_{k2}	…	c_{kn}	X_{k1}	X_{k2}	…	X_{kn}

2.2.3 Estimating the unknown data

1. The steady-state and unsteady-state flow

In fluid mechanics, flow can be divided into two types: the steady-state and unsteady-state flow. A steady flow is one in which the conditions (velocity, pressure and cross-section) may differ from point to point but DO NOT change with time. If at any point in the fluid, the conditions change with time, the flow is described as unsteady.

The flows and stocks of the identified systems are determined by measurements, market research, expert judgment, best estimates, interviews, and "hands on" knowledge. The MFAs generally assume that in the production, manufacturing, and waste management processes, no material is stored or the net flow during the sample time is zero, that is, that this part of the system can be treated as static. Hence, the dynamic modeling approaches focus on the use phase (which has nonzero net flows) and the resulting in-use stock changes. Stocks and flows are often modeled as time series with a constant sampling rate t, that is, $f[n]=f(nt)$, typically with $t=1$ year. Material balances are performed on those processes where no data are available (using the principle of mass conservation: $\text{inflow}(t)=\text{outflow}(t)$). If required, results can be integrated into static or dynamic models. Modelling different scenarios is useful to assess the impact of various measures on the regional stocks and flows of selected materials in view of environmental loads or resource depletion.

In mathematical terms, the steady-state calculation can be described as follows. Consider a single node with an input flow, a stock $S(t)$, and an output flow $O(t)$. $O(t)$ is proportional to S ($O=kS(t)$), where k is a rate constant that can be considered as the reciprocal lifetime (of the substance in the node). As $O(t)$ is not necessarily equal to inflow (t), $S(t)$ will change over time. After writing the mass balance as a first-order differential equation:

$$\frac{dS}{dt}=\text{inflow}(t)-kS(t) \tag{2-4}$$

and integrating, if follows, assuming $\text{inflow}(t)$ is constant, that

$$S(t)=e^{-kt}\left(S(0)-\frac{\text{inflow}}{k}\right)+\frac{\text{inflow}}{k} \tag{2-5}$$

where $S(0)$ is the initial stock.

After a long or even infinite time, S will no longer change: an equilibrium in reached whereby $S(\infty)=\text{inflow}/k$. this equilibrium is called the steady state. For complex systems it is similarly possible to determine the behavior of stocks and flows over time given a set of constant inputs. But in most cases, MFA belongs to the unsteady-state flow. For instance, we consume the electronics and discard the e-waste every year (Figure 2 - 6). Owing to a certain of duration of use process, the e-waste amount in same year does not generally equal the electronics amount.

In mathematical terms, the unsteady-state calculation can be described as follows:

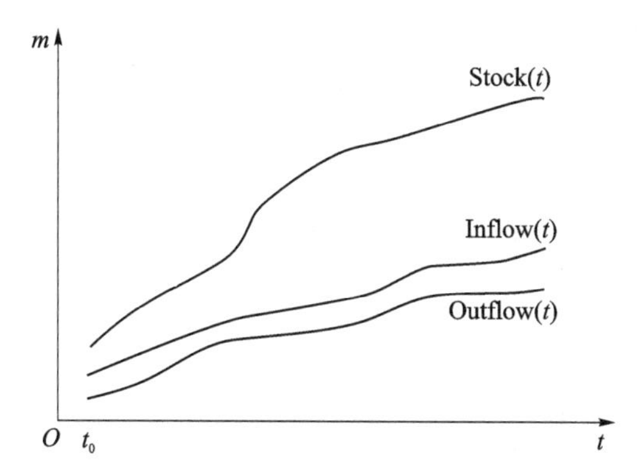

Figure 2 - 6 The functions of inflow, outflow, and stock of material flow in an unsteady state

$$\text{Stock}(t) = \int_{t_0}^{t} \text{inflow}(t)\,dt - \int_{t_0}^{t} \text{outflow}(t)\,dt + \text{Stock}(t_0) \tag{2-6}$$

where stock(t) is the cumulative stock at the time t; $\int_{t_0}^{t} \text{inflow}(t)\,dt$ is the cumulative input at the time t;

$\int_{t_0}^{t} \text{outflow}(t)\,dt$ is the cumulative outflow at the time t; stock(t_0) is the stock at the time t_0.

2. The top-down and the bottom-up approach

The material stock of a process can be measured by two different methods. The first method, usually referred to as the top-down approach, derives the stock from the net flow: the difference between inflows and outflows. The second method, the bottom-up approach, directly estimates the stock by summing up the material in question present within the system boundary at a certain time. For instance, there is 20 Mt copper consumption in 2020. Copper is used in various products. From the top-down approach, copper consumption in 2020 was 20 Mt; but from the bottom-up approach, the total copper consumption in 2020 would be the sum of various products like mobile phone, computer, TV, air conditioner, vehicle, camera, washing machine, and airplane, etc (Figure 2 - 7).

The top-down approach derives the in-use stock S from the net flow by using the balance of masses as given in Eq. (2 - 7).

$$dS(t) = (\text{inflow}(t) - \text{outflow}(t)) \cdot dt = \text{netflow}(t) \cdot dt \tag{2-7a}$$

$$S[n] = (\text{inflow}[n] - \text{outflow}[n]) \cdot T + S[n-1] \tag{2-7b}$$

$$S[N] = T \cdot \sum_{n=1}^{N} (\text{inflow}[n] - \text{outflow}[n]) + S[0] \tag{2-7c}$$

The second method, referred to as the bottom-up approach, derives the in-use stock $S[n]$ at a time n by summing all the material contents c_i in their respective products or end-use sectors P_i with the following equation:

$$S[n] = \sum_{i=1}^{I} P_i[n] \cdot c_i[n] \tag{2-8}$$

where I is the total number of products or end-use sectors considered. To construct time

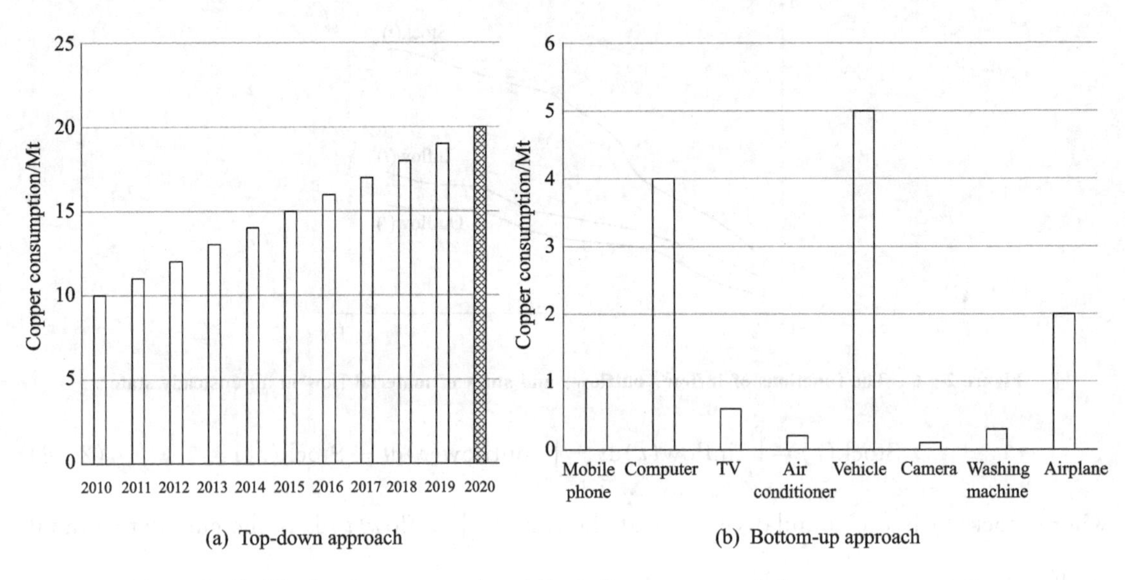

Figure 2 - 7 The top-down approach and the bottom-up approach for copper consumption

series of in-use stock, $S[n]$ is determined for every requested year n. If required, net flow can be calculated by introducing Eq. (2 - 8) into Eq. (2 - 7b). In fact, the majority of previous works used the top-down approach, rather than the bottom-up approach.

For inflows, historical data (e. g. on a long-term consumption) is often accessible, but outflows are rarely measured. Outflows are quantified commonly by assigning lifetime distribution functions to specific products or end-use sectors, with the relationship between inflows and outflows corresponding to a convolution (Eq. (2 - 9) with $*$ denoting the convolution; this approach is also called the residence time model or population balance model). Since it is rarely possible to solve this convolution analytically, it is integrated numerically according to Eq. (2 - 10).

$$\text{outflow}(t) = (\text{inflow} * f)(t) = \int_{-\infty}^{\infty} \text{inflow}(t-u) \cdot f(u)\,\mathrm{d}u \tag{2-9}$$

$$\text{outflow}[n] = \sum_{m=-\infty}^{\infty} \text{inflow}[n-m] \cdot f[m] \tag{2-10}$$

where $f(t)$ and $f[m]$ are the probability densities of the lifetime distribution function for the continuous and the time discrete case, respectively.

2.2.4 MFA diagram and its interpretation

When all the data, collected and estimated, is enabled for each flow and stock. The system can be visualized with an MFA diagram. For instance, making a cup of coffee is illustrated from materials input to product output in an MFA diagram (Figure 2 - 8). 10 g coffee beans, 2 g filter, and 250 g water are processed to make coffee, and finally 220 g for coffee product, 41 g for used filter and coffee ground, and 1 g residue powder.

The "interpretation" of the results, which comprises an evaluation in terms of soundness, robustness, consistency, completeness, etc, as well as the formulation of

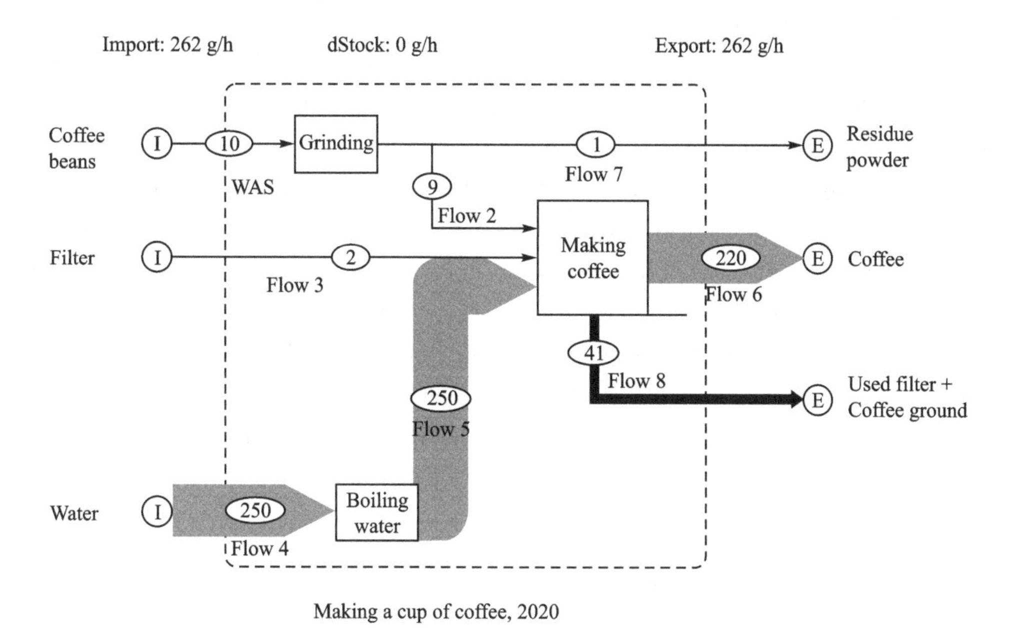

Figure 2 - 8　MFA diagram for making a cup of coffee

conclusions and recommendations. The results of MFA studies are interpreted taking into consideration loading quantities, the significance of stocks and the comparisons of results against environmental standards and/or sustainable indicators or other assessment approaches. Again, as the given case of making a cup of coffee (Figure 2 - 8), some interpreted and extended questions can be proposed:

① How many waste generations in making a cup of coffee?

② What is the use efficiency of coffee beans in making a cup of coffee?

③ What is the total resource efficiency in this process?

In order to make effective use of the MFA results in decision-making, the interpreted results need to be communicated to the relevant policy makers and stakeholders (water, energy, waste, transport and environment management bodies, community groups, non-governmental organizations, and representatives from neighboring regions). Workshops may be conducted with relevant stakeholders to determine: what these results mean for each stakeholder and management group in the region and how the region wants to respond to these findings; what priorities these material flow and stock issues should take in the region's current and future development strategies; and what the implications of these flow/stocks are for other policies/projects/programs.

2.3　Uncertainty and sensitivity analysis

2.3.1　Data error and distribution

Uncertainty or error in flow and stock estimates can be grouped into two categories:

modeling identification and data uncertainty (parameters) (Figure 2 - 9). The latter can be also classified into two types: aleatory variability and epistemic uncertainty. Aleatory variability is the natural randomness in a process. Regarding discrete variables, the randomness is parameterized by the probability of each possible value. Regarding continuous variables, the randomness is parameterized by the probability density function. Epistemic uncertainty is the scientific uncertainty given in the model of the process. This is due to limited data and knowledge. The epistemic uncertainty is characterized by alternative models. For discrete random variables, the epistemic uncertainty is modelled by alternative probability distributions. For continuous random variables, the epistemic uncertainty is modelled by alternative probability density functions. Moreover, there is epistemic uncertainty in parameters that are not random by having only a single correct (but unknown) value.

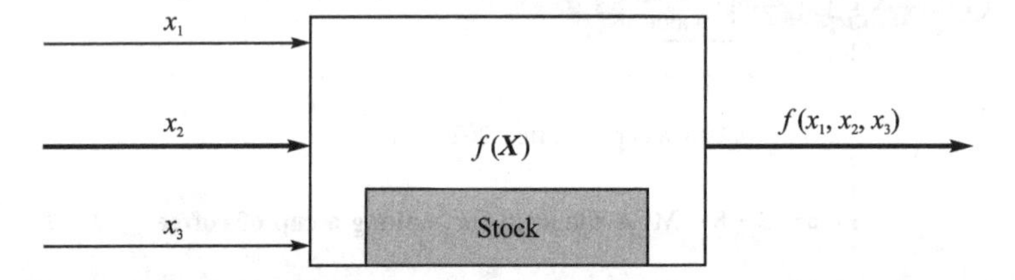

Figure 2 - 9 Uncertainties source and propagation from aleatory variability and epistemic uncertainty

Aleatory variability arises due to inherent variability, natural stochasticity, environmental or structural variation across space or through time, manufacturing or genetic heterogeneity among components or individuals, and a variety of others sources of randomness. Aleatory variability cannot be reduced but only better identification. On the other hand, epistemic uncertainty arises due to insufficient knowledge about the real world, which are deprived from small sample sizes, detection limits, and imperfections in scientific understanding. Epistemic uncertainty can be reduced by further investigation.

While aleatory variability can be handled by probability theory, epistemic uncertainty is better covered by possibility or fuzzy set theory. Independent of this fact, epistemic errors are often treated in terms of probabilities as if they were aleatory. Consequently, any representation requires subjective assumptions. In the following, the probabilistic approach is used for both kinds of uncertainties. Even though, in uncertainty analysis, epistemic uncertainties (e. g. measurement errors) are frequently assumed to be normally distributed, in scientific models in general and in MFA models, this is often not the case. If, for instance, a process model is correct, mass flows and concentrations cannot take negative values, and transfer coefficients are restricted to the unit interval. However, if uncertainties are assumed to be normally distributed but are not, such negative values might occur. Another example is provided by expert opinions that frequently must be relied on in MFA due to scarce or missing data. They are often modeled by a uniform, triangular, or trapezoidal distribution depending on the level of knowledge of the expert.

2.3.2 The identified approaches for uncertainty analysis

The evaluation of uncertainties in an MFA involves four elements: evaluation of uncertainties in the input to each of the tasks of an MFA; propagation of input uncertainties through each task; combination of the uncertainties in the output from the various tasks; and display and interpretation of the uncertainties in the MFA results. Table 2 - 2 illustrates the identified approaches to reduce the errors from modeling and parameter to completeness.

Table 2 - 2 Source and types of uncertainties and those identified approaches

Category	Main identified approaches
Modeling	Is the model adequate? For example, do the binary event-tree and fault-tree models represent the continuous process adequately? Is uncertainty introduced by the mathematical or numerical approximations that are made for convenience? If the model is valid over a certain range, is it being used outside that range?
Parameter	Data may be incomplete or biased. In licensee event reports, for example, are we sure that all relevant failures are listed, and do we know the number of trials? Do the available data apply to the particular case? This raises the question of generic vs. site-specific data. Is the method of data analysis valid? Do the data really apply to the situation being studied? For instance, are all pumps in all plants in the data base expected to have the same failure rate, or should they be regarded as variable?
Completeness	Have the analyses been taken to sufficient depth? Have all important processes been treated?

The most used measures of uncertainty are probabilistic statements about the values of parameters, but the concept of probability is interpreted differently by classical and Bayesian analysts. In theory, uncertainty analysis related to input data error can be explained by Eq. (2 - 11):

$$\Delta f=\sqrt{\left(\frac{\delta f}{\delta x_1}\Delta x_1\right)^2+\left(\frac{\delta f}{\delta x_2}\Delta x_2\right)^2+\left(\frac{\delta f}{\delta x_3}\Delta x_3\right)^2} \tag{2-11}$$

where $\frac{\delta f}{\delta x}$ is a partial derivative, Δx is the changing value of input data, Δf is the changing value of output data, and 3 is the three-input flow indicated from Figure 2 - 9.

Inflow data distribution, like $\boldsymbol{X}$ given in Figure 2 - 9, is the fundamental information to enable the final outflow and stock. The common data distributions include Beta, Burr, Cauchy, Chi-Squared, F, Fatigue life, Gamma, Normal, and Weibull distribution, whose probability density functions are given here in Figure 2 - 10.

2.3.3 Mathematic methods for uncertainty analysis

Two straightforward methods to propagate uncertainties are frequently employed in MFA.

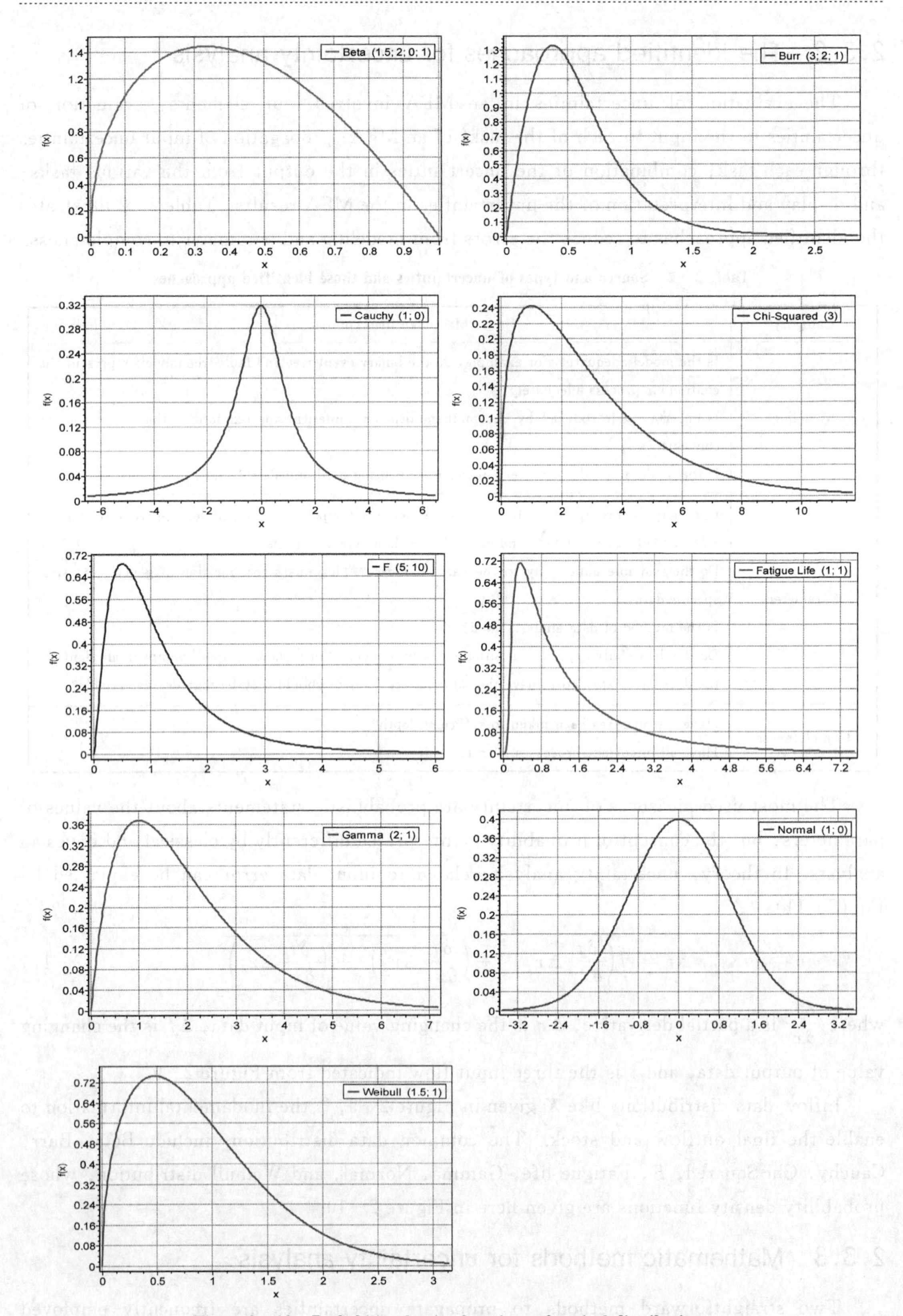

Figure 2 - 10 Probability density functions of typical distributions

1. Gauss's Law

The propagation of uncertainties can be evaluated by applying Gauss's law of error propagation to a function of interest.

$$Y = f(x_1, x_2, x_3, \cdots, x_n) \tag{2-12}$$

where Y is the outflow or stock, x is the inflow, f is the function in the process.

The expectation of the result Y can be estimated from Eq. (2-12) and the variance of the result from Eq. (2-13). The latter is called Gauss's law of error propagation.

$$E(Y) \approx f(E(x_1), E(x_2), E(x_3), \cdots, E(x_n)) = f(\mu_1, \mu_2, \mu_3, \cdots, \mu_n) \tag{2-13}$$

$$\mathrm{var}(Y) \approx \sum_{i=1}^{n}\left(\mathrm{var}(x_i) \cdot \left[\frac{\delta Y}{\delta x_i}\right]^2_{x=\mu}\right) + 2\sum_{i=1}^{n-1}\sum_{j=i+1}^{n}\left(\mathrm{cov}(x_i, x_j) \cdot \left[\frac{\delta Y}{\delta x_i}\right]_{x=\mu} \cdot \left[\frac{\delta Y}{\delta x_j}\right]_{x=\mu}\right) \tag{2-14}$$

where μ is the expectation or mean of x, $\mathrm{var}(Y)$ is the variance of Y, and $\mathrm{cov}(x_i, x_j)$ is the covariance of x_i and x_j.

2. Monte Carlo Simulation

Monte Carlo simulations (MCS) are used to simulate the probability of different outcomes in a process that cannot be easily predicted due to random variable interventions. It is a technique used to understand the effects of risk and uncertainty. MCS are used to solve problems in many areas, including investing, commerce, physics, and engineering. Unlike normal forecasting model, MCS predicts the results of a set of values based on an estimated range versus a set of fixed input values. In other words, MCS build a model of possible outcomes for any variable with inherent uncertainty by using probability distributions, such as uniform or normal distributions. It then recalculates the results again and again, using a different set of random numbers between the minimum and maximum values each time. In a typical Monte Carlo experiment, the process can be repeated thousands of times to produce a large number of possible results.

Regardless of what tool you use, Monte Carlo techniques involves three basic steps:

① Set up the predictive model, identifying both the dependent variable to be predicted and the independent variables (also known as the input, risk or predictor variables) that will drive the prediction.

② Specify probability distributions of the independent variables. Use historical data and/or the analyst's subjective judgment to define a range of likely values and assign probability weights for each.

③ Run simulations repeatedly, generating random values of the independent variables. Do this until enough results are gathered to make up a representative sample of the near infinite number of possible combinations.

Let's take a look at an example in order to better understand the concept behind the MCS. For that, we firstly consider the simple model representation of an MFA process, as shown in Figure 2-11.

Basically, a process consists of a series of stages designed to produce a product and/or a service as required by the user. A process is mainly composed of three parts: Inflow(x), process steps $f(x)$ and outflow(y). In our example shown in the Figure 2 - 11, we have some process inflow (x_1, x_2, x_3, $\cdots$, x_n), one process outflow (y) and three process steps. The process inputs can be understood as the materials we put into the process aiming at achieving a specific outflow(y), whereas the process outflow(y) can represent, for instance, the final product delivered to the user.

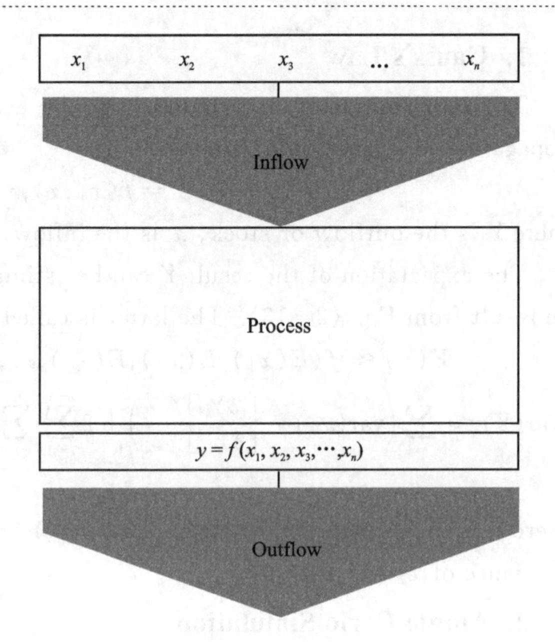

Figure 2 - 11 MFA from inflow to outflow via process

Any change of data distribution in the inputs of the process causes variation in the outflow(y). Therefore $y=f(x)$, in other words, the process outflow(y) is a function of the process inflow(x). Generally speaking, if we keep the process inputs always the same, the process output is also expected to be the same (i. e. small output variations) and can be evaluated straightforwardly. Put in other words, equal initial inputs will produce a similar final result, where the final result can be a product or service. Let's say now that each process inputs vary in a defined range of probable values. Thus, in this case, it is reasonable to expect that the process output will vary as well. Additionally, as long as we add more and more varying inputs to the process, it will get complex to evaluate the process output. For such complex processes you should consider using the MCS in order to help you predict the process output.

Example 2 - 1

Take the example of end-of-life vehicles (ELVs) in Austria. The number of ELVs in 2010 is reported to be some 245,000 vehicles for that year, and the mean copper content of a car is assumed to be 20 kg. This would result in a yearly copper flow of 4,900 tons. Because it is not obvious whether this number has been rounded to two, three, or four significant figures when quoted without the calculation, it is better to state the result in scientific notation as 4. 9 kilotons (kt) or, even better, 4. 9 kt (but this is another story…) to emphasize the two-significant-figures concept.

In our example of ELVs, we assume that the number of ELVs has an SD of 20,000 ELV (= 8%) and the mean copper content per vehicle an even higher SD of 4 kg (= 20%). Whereas the first is an educated guess, the latter could be based on results of various studies or values found in the literature. By applying Gauss' law of error propagation, the uncertainty of the already known copper flow of 4. 9 kt can be computed, resulting in an approximate SD of 1 058 tons (= 22%).

If we assume the result of our example to be normally distributed and apply the rounding rules, we get (4.9±1.1) kt of copper with a 95% uncertainty range (=MV± 1.96 · SD) from 2.8 to 7.0 kt (all data rounded after calculation). Because the normal distribution extends from minus to plus infinity, there is always a chance to get negative values. To avoid such unrealistic negative values for mass flows and concentrations, it helps to reject the normal distribution assumption and apply distribution functions that exclude negative values (e.g. lognormal). In this case, MCS can be applied to calculate the uncertainty ranges of flows and stocks. Tools for running MCS in Microsoft Excel are available as freeware (e.g. Monte Carlo) or commercial software (e.g. Crystal Ball) and can be easily applied. Gauss's law of error propagation can be well applied for bottom-up stock calculations when the final result, as in the case of copper, is derived from proxy or intermediate data, such as the number of vehicles, kilometers of railways, square meters of roofs, and so on. Even if the single proxy data might have considerable uncertainty ranges, this need not result in unacceptable uncertainty ranges of the final result (say, SD >50%), because random errors do not add up, but rather tend to compensate each other to a certain extent.

2.3.4 Sensitivity analysis

In fact, there is, in most cases, no adequate information to fulfill the uncertainty analysis. Sensitivity analysis is the study of how the uncertainty in the output of a mathematical model or system (numerical or otherwise) can be divided and allocated to different sources of uncertainty in its inputs. A related practice is uncertainty analysis, which has a greater focus on uncertainty quantification and propagation of uncertainty; ideally, sensitivity analysis should be run in random. Sensitivity analysis helps determine how changes in one input affect the output.

Deprived from Eq. (2-12), its sensitivity or propagation of error considerations could be determined as

$$\frac{\Delta Y}{Y}=\frac{f((x_1+\Delta x_1),(x_2+\Delta x_2),(x_3+\Delta x_3),\cdots,(x_n+\Delta x_n))-f(x_1,x_2,x_3,\cdots,x_n)}{f(x_1,x_2,x_3,\cdots,x_n)} \tag{2-15}$$

where $\frac{\Delta Y}{Y}$ is the sensitivity, and Δx is the range of inflow.

2.4 Main software of material flow analysis

MFA has absorbed a wide concern from the academic to industry society. Many software has been fostered in recent years. Some famous software includes STAN and e! Sankey. Microsoft Power BI, OriginLab OriginPro, and Microsoft Excel have the new function of Sankey to map the MFA. Additionally, some websites have the simple and fast expression for MFA.

2.4.1 STAN

STAN (short for subSTance flow ANalysis), a free software developed at the Vienna University of Technology, has been written for the specific purpose of performing material flow analyses. It is freeware, but the source code is unavailable to the public. STAN is a stand-alone software for Windows. Its publisher is TU Wien, Institute for Water Quality, Resources and Waste Management, Department of Waste and Resources Management.

STAN software Download Instructions are given as follows:

① Go to http://www. stan2web. net.

② Click "Register" (on the right side, under Login window).

③ Enter your registration data, type in security code and click the "Register" button.

④ You will receive an automatically sent registration email. Confirm your email address by clicking the hyperlink within.

⑤ Your account is now activated. Login with your username and password.

⑥ In the menu select "Downloads > STAN".

⑦ Download the latest STAN version.

⑧ Unpack the downloaded zip-File.

⑨ Click "setup. exe" and follow the installation instructions for the Windows operation system.

After installation, its interface is looking as Figure 2 - 12. STAN Help in the software provides the detail information about how to use it. In STAN a context sensitive help is

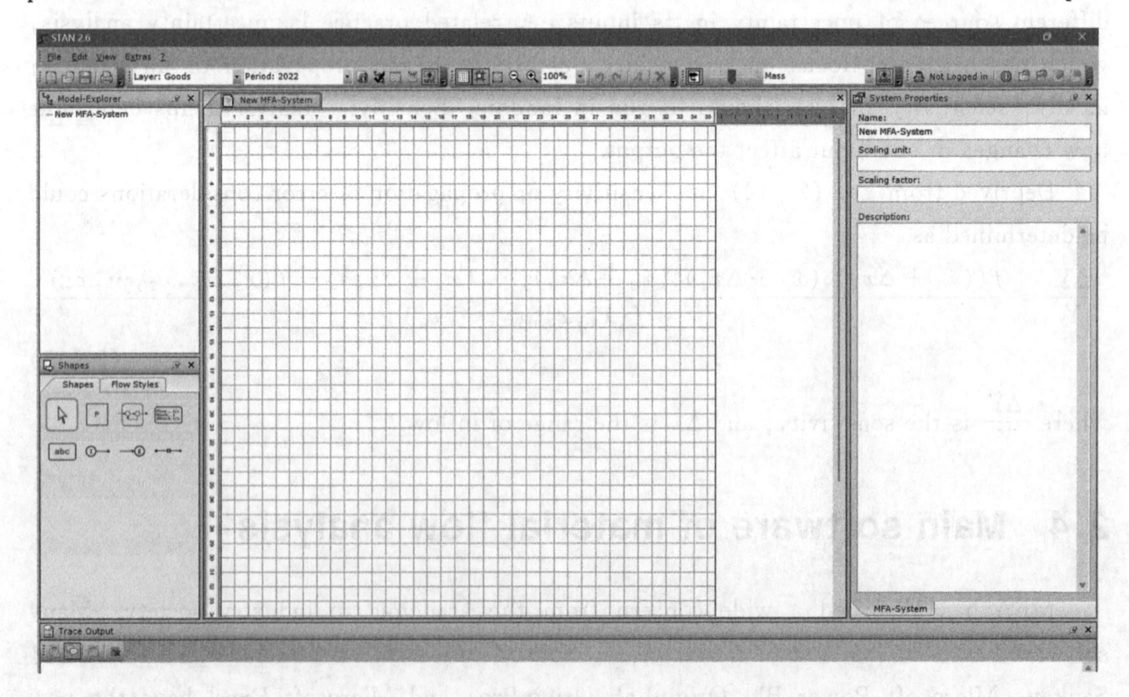

Figure 2 - 12 STAN software operation interface

implemented which can be activated by pressing F1 on the key board. Besides general information about the active window, links to certain topics of the manual can be found. The language of the user interface including the help file can be switched between English and German. The main idea behind STAN is the combination of all necessary features of a MFA in one software product: graphical modelling, data management, calculations and graphical presentation of the results.

STAN not only is free software, but has some specific merits. For instance, STAN allows the consideration of data uncertainties. It is assumed that uncertain quantities are normally distributed, given by their mean value and standard deviation. This approximation offers the possibility to use methods like error propagation and data reconciliation. If an unknown quantity is to be calculated from a function of independent random variables, its uncertainty (variance) can be calculated by the law of error propagation. Data reconciliation can be enabled in an automatic way by this software. During data reconciliation, the mean values of uncertain data will be altered in a way that contradictions disappear. A solution is found when the sum of squares of the necessary chances reaches a minimum (method of least squares). The inverse of the variances (square of standard uncertainty) of the uncertain quantities are used as weighing factors. As additional effect of data reconciliation the uncertainty of the reconciled data is reduced.

2.4.2 e! Sankey

e! Sankey is the standard Sankey diagram software for professionals (https://www.ifu.com/e-sankey/). You can easily create Sankey diagrams with high accuracy and neat layout. Diagrams are versatile in use, no matter whether you wish to create an energy flow, material flow, cost flow or a supply chain/process flow diagram. Use e! Sankey to create convincing visualizations for your presentations, reports and publications and link them to your Microsoft Excel files. The perfect visualization for resource and energy efficiency: a Sankey diagram immediately draws the attention to the most important flows.

e! Sankey can beadopted in many different fields of applications, such as:

① material flows and efficiency.

② logistics, transport of goods, supply chain.

③ waste water and waste disposal.

④ energy audits and energy management.

⑤ energy flows, balance, and efficiency.

⑥ heat transfer and heat losses.

⑦ technical processes, chemical engineering.

⑧ visualization of cost flows & value streams.

2.4.3 A web-based approach

While the existing software products offer the necessary functionality to perform MFAs, they exhibit certain drawbacks: Most of them are only available for Windows and they all act as stand-alone applications to be installed on a client computer. All participants in a research project, as well as people interested only in the results, need to own the software and install it on their machines.

There is a web-based approach to enable MFA. Here are the two examples as noted here.

① https://online.visual-paradigm.com/cn/charts/explore/sankey-diagram-maker/(Figure 2-13).

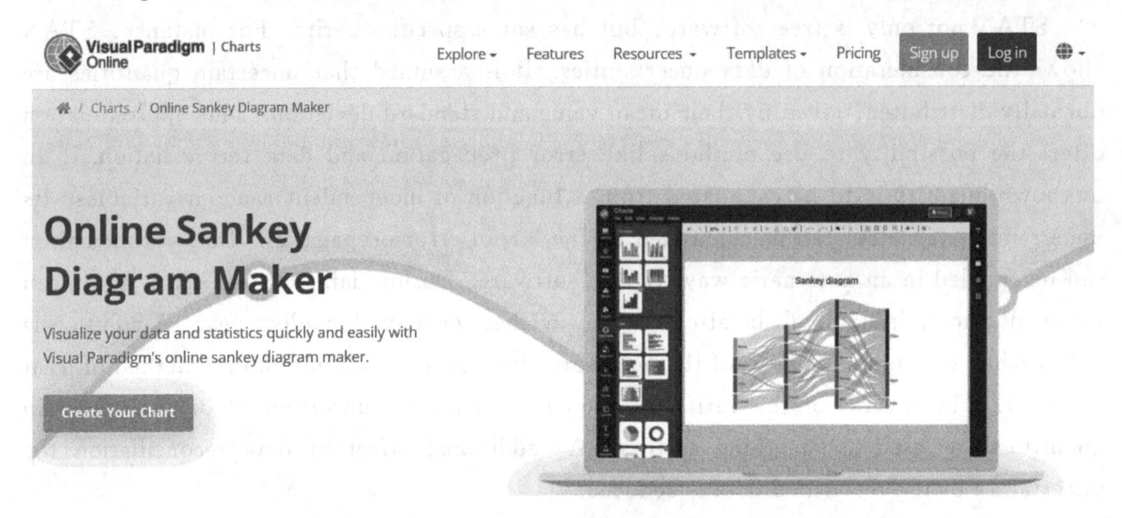

Figure 2-13 Online Sankey Diagram Maker

② https://sankeydiagram.net/(Figure 2-14).

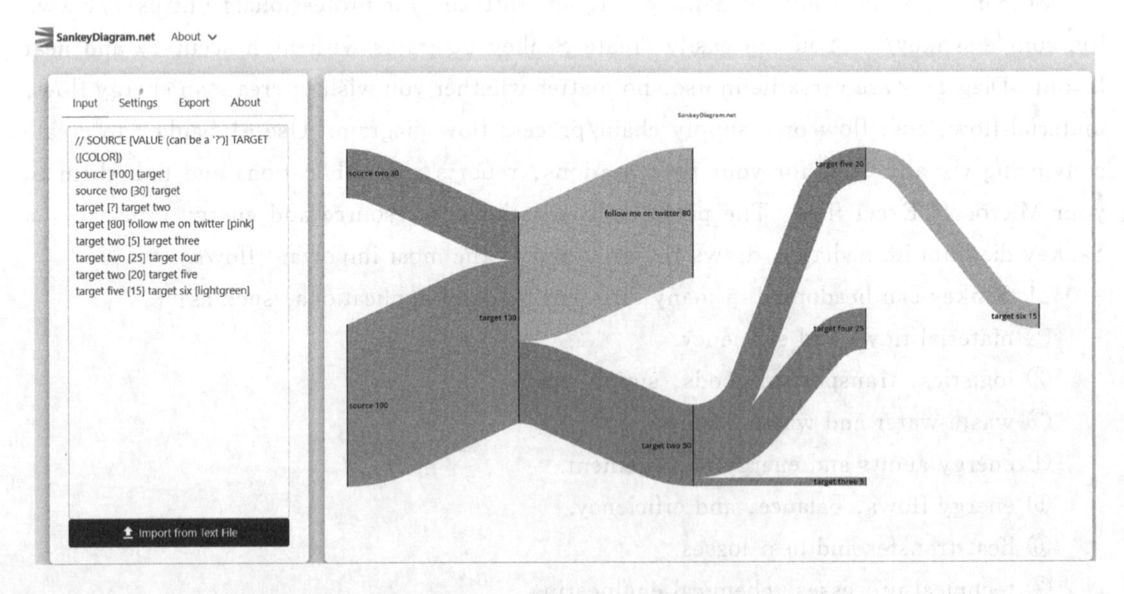

Figure 2-14 Online Sankey Diagram

2.5 Further reading

[1] Brunner P H, Rechberger H. Handbook of Material Flow Analysis: For Environ-

mental, Resource, and Waste Engineers[M]. 2nd ed. Baca Raton: CRC Press, 2016.
[2] Muller E, Hilty L M, Widmer R, et al. Modeling metal stocks and flows: a review of dynamic material flow analysis methods[J]. Environmental Science & Technology, 2014, 48 (4): 2102-2113.
[3] Chen W Q, Graedel T E. In-use product stocks link manufactured capital to natural capital[J]. Proceedings of the National Academy of Science of the United States of America, 2015, 112 (20): 6265-6270.
[4] Cencic O, Rechberger H. Material flow analysis with software STAN[J]. Journal of Environmental Engineering Management, 2008, 18 (1): 3-7.
[5] Rechberger H, Cencic O, Frühwirth R. Uncertainty in Material Flow Analysis[J]. Journal of Industrial Ecology, 2014, 18(2): 159-160.
[6] Christensen P, Gillingham K, Nordhaus W. Uncertainty in forecasts of long-run economic growth[J]. Proceedings of the National Academy of Sciences, 2018, 115 (21): 5409-5414.
[7] Douglas-Smith D, Iwanaga T, Croke B F W, et al. Certain trends in uncertainty and sensitivity analysis: An overview of software tools and techniques[J]. Environ Modell Software, 2020, 124: 104588.
[8] Miatto A, Wolfram P, Reck B K, et al. Uncertain future of American lithium: a perspective until 2050[J]. Environmental Science & Technology, 2021, 55 (23): 16184-16194.

2.6 Exercises

1. What is the relationship among three laws of anthropogenic circularity science?

2. China's copper consumption in 2009, 2010, and 2011 was estimated as 4.2, 4.5, and 4.8 million tons, 10% of which was used in the electrical and electronic industry. The lifespan of electrical and electronic equipment ranges from 8 to 12 years. On the other hands, China's reported e-waste amount in 2020 was 13 million tons, consisting of 6 million tons electronic waste and 7 million tons electrical waste. Electronic waste and electrical waste contain the 5% and 2% copper composition, respectively. Please use the top-down and the bottom-up approach to estimate copper amount contained in the e-waste.

3. In most cases the sum (Σ) is commonly used in MFA. How to fill the gap from Σ to infinitesimal calculus ($\int$) while estimating the unknown flow and stock? Or how to transplant the infinitesimal calculus in MFA?

4. What are differences between uncertainty and sensitivity analysis? Please use one example to illustrate.

5. Select one product of moderate complexity, such as a computer or smart phone. Conduct an MFA on the product along all life cycle. Prepare a report that summarizes your findings, comment on where itwas difficult to assign rating because of lack of information, and propose design changes that would improve the environmental responsibility of the

product.

6. Take an environmental or resource report of a regional company or sector of your choice (e. g. pulp and paper mill, electricity supplier, cement manufacturer, car manufacturer, food producer; many reports can be downloaded from the World Wide Web) and analyze it from the point of view of material balances. What are the most important material and substances (quantitatively and qualitatively)? Is it possible to establish material balances based on the information provided? If not, which data are missing? Find issues for improvement by MFA.

7. Please download one recent publication with the e! Sankey diagram and map it using the STAN again.

Chapter 3 Applications and Case Studies

MFA has been employed to answer some multidisplinary questions from national science and engineering, to economic and management insight. The three main and specific areas for MFA application summarized are resource management, environmental management, and waste management. In addition, industrial ecology, anthropogenic metabolism, green consumption, and social economy are overarching themes where MFA plays a key role. Today's emerging fields for MFA are manufacturing, process design, and engineering. In all of these fields, MFA has been used for analysis and optimization on all scales. The specific tasks tackled by MFA comprise the improvement of resource efficiency; the design of measures for urban mining; pollution prevention and protection of, for example, surface water and groundwater; watershed and nutrient management; waste minimization; waste analysis; recycling; and the provision of final sinks.

3.1 Resource management

3.1.1 Material flow analysis for metals

Our modern society heavy relies on types of minerals, which are indispensable to maintain the quality of life and support the increasing economic growth. On the one hand, the continuing growth of demand is depleting the finite mineral resources and causing irreversible impacts on the planet. The current oversupply of raw materials in world markets masks a persistent underlying global challenge. On the other hand, increased consumption has led to an accumulation of significant stocks of metals in the anthroposphere, and the collection and recycling of metals from these secondary resources has become more and more important. These activities rely on mapping the anthropogenic material cycles regarding sources, quantities, and locations of metal-containing goods that have accumulated in the past.

MFA has been widely used to track the fate of metals along their whole life cycles in a certain system. MFA studies are abundant on tracing the critical materials stocks and flows with national and global boundaries, such as cobalt, molybdenum, lithium, rare earth elements, indium, tantalum, gallium, and platinum.

1. For bulk metals

(1) Case study: the global cycle of steel during 1900—2015

Steel is the most used metal in our modern society. However, steel production is highly energy- and carbon-intensive and considered a difficult-to-mitigate sector. Strategies for de-carbonizing the steel industry have primarily focused on production efficiency improvement, including energy efficiency measures, production technologies innovation and fuel switching. However, the effectiveness of such production-based strategies in terms of carbon reduction has recently been questioned. These calls for attention to gauge the entire progress that the global steel industry has made on GHG mitigation. Most of previous investigations have been limited to specific production technologies where the interplay between material flows and supply-side technical efficiency was widely overlooked.

This case integrates dynamic MFA with LCA to estimate annual production, efficiency and GHG emissions of global steel production based on dominant processes during 1900—2015 (Figure 3 - 1). By examining the interplay between material flows and GHG emissions, the results shows that: ① The steel industry consumed ~46 Gt iron ore and ~31 Gt home, new and old scrap to produce ~45 Gt steel products during this period, which, in terms of weight, is around one-third of the total steel production-related GHG emissions (147 Gt CO_2-eq); ② At present, over half of those steel products remain as societal in-use stocks (i. e. ~25 Gt) with the largest share stored in buildings (~16 Gt) which were mainly constructed in the past two decades (driven by large emerging economies, including China and India). In general, those societal in-use stocks are quite young with ~83% of global steel in-use stocks being built after 1990; ③ Given that the average lifetime of steel products is ~70 years, a rapid increase in old scrap generation can be foreseen in these countries over the next 30~50 years. Indeed, the past few decades have already witnessed a significant increase in old scrap generation from ~45 Mt/year in 1950 to ~427 Mt/year in 2015, concomitant with a remarkable improvement in steel recycling rate (now remaining at around 70%); ④ In addition, the result of decomposition analysis reveals the contribution of efficiency improvement and production outputs to these emission changes, the results of which highlighted the inadequacy of process efficiency alone in achieving absolute emissions reduction. The case indicates that the GHG intensity of the global steel industry had stagnated in the past 15~20 years before 2015. This stagnation indicates the urgency of the joint implementation of process efficiency and demand-side measures to reduce GHG emissions and achieve climate targets.

(2) Case study: the copper's stocks and flows in China

Copper is one of the most pervasive materials with high corrosion resistance, high ductility, and good electrical conductivity. As the largest developing country, China has become the world's largest copper producer and consumer. However, the issue of supply security of copper consumption in China would significantly constrain the sustainable development of China's copper industry, even affecting the promotion of a circular economy. Owing to a large copper consumption and poor copper reserves, reducing the support for copper resources' economic development is pivotal to China's sustainable development. But as the world's factory, China also exports a lot of resources and products to other countries. The important role of trade in such metabolism has been limited explored and appreciated.

This case develops China's copper cycle, and analysis uncover trade importance in national metabolism from 1950 to 2015 to explore a high-resolution China's copper cycle in

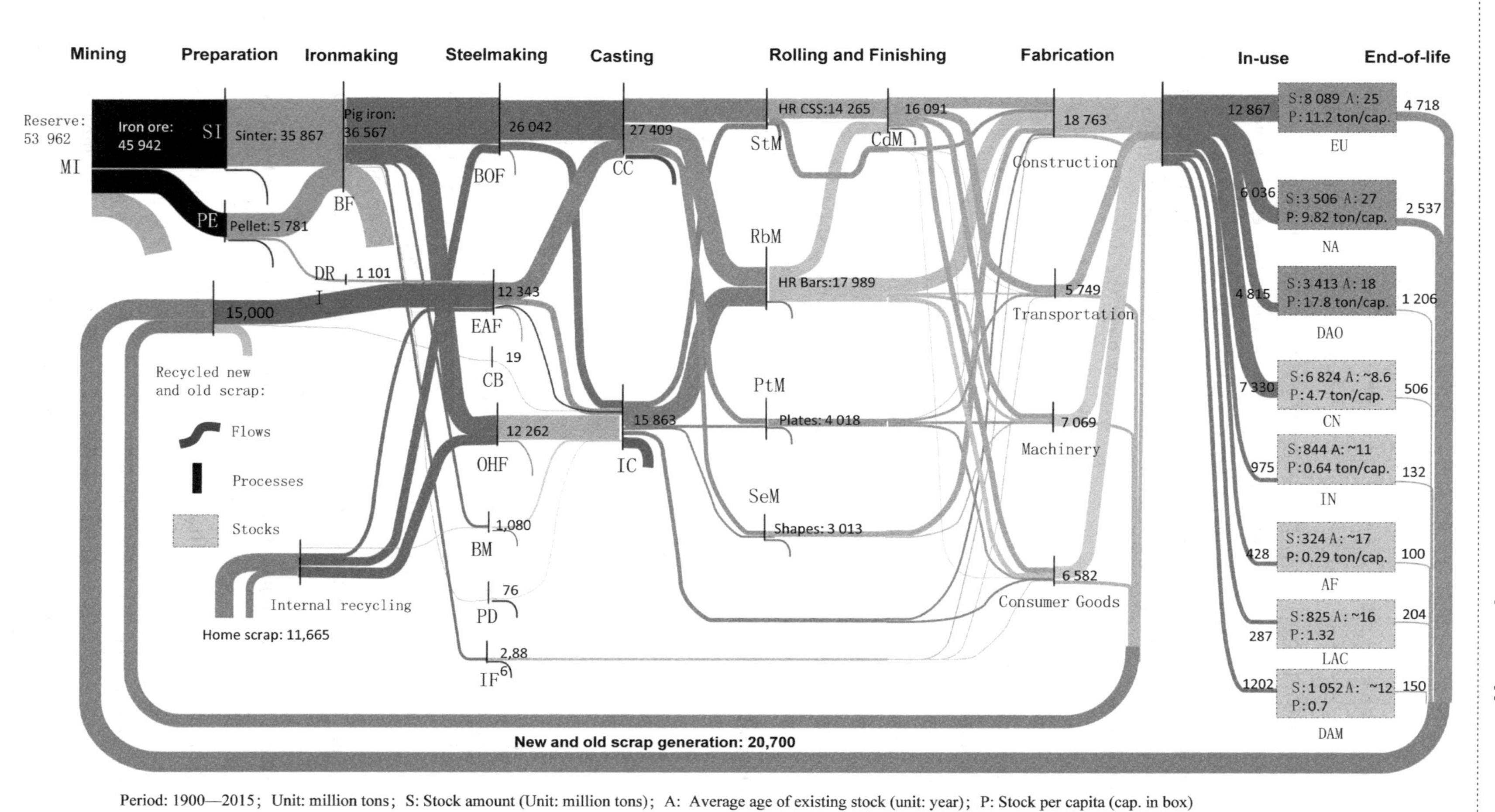

Period: 1900—2015; Unit: million tons; S: Stock amount (Unit: million tons); A: Average age of existing stock (unit: year); P: Stock per capita (cap. in box)

Figure 3–1 The global historical steel cycle in a Sankey diagram where the numbers represent the accumulated annual flows over the past 115 years

the socioeconomic system (Figure 3 - 2), identify the role of products of copper contained trade uncovers in national metabolism, and help to eliminate the external dependence of copper resources and realize the sustainable development of copper resources. Around 70% of China's copper use relies on other nations, with around half of those being contributed by the copper waste trade. In particular, the copper waste import contributed around 28 million tons of copper ore reductions, equivalent to the entire copper reserves in China. Consequently, more attention should be paid to the importance of the metal waste trade since it may result in increased mineral dependence and extraction, shift the environmental burden to primary copper resources, and potentially threaten secondary copper markets in China. Domestic production of copper waste in China can fully meet the demand for copper, even if there is no import of copper scrap based on the copper cycle in Chinese metabolism. However, China's waste management system of copper is far from recycling EoL copper efficiently, which highlights the need for policymakers to build an effective recycling system and improve the utilization rate of copper scrap flow into the manufacturing process. The high-resolution investigation of copper flows reveals that China dominates the global copper market and copper resource trade accounting for 50% of the world's copper consumption and has become a significant factor affecting the global copper demand. Furthermore, China's copper consumption is enabled by imports of copper-containing products, and such high dependence on international trade is becoming a pressing challenge to its resource security.

2. For critical metals

The demand for natural resources is rising to an unprecedented level, especially for the critical materials that are problematic in supply but economic important. Besides, the demand for critical materials is projected to surge over the following decades, as these materials are used in many environmentally friendly technologies. However, most critical materials are uneven geographical distribution, and such contradiction between limited resource supply and surging demand has induced concerns about material criticality and supply chain disruptions. Several studies have established ratings of material criticality by calculating some indicators. The U. S. National Research Council (2007) conducted static matrix to assess material criticality, which has been widely adopted in many criticality studies. However, criticality is a dynamic state relates to a mineral's whole supply system, which calls for an in-depth analysis of material uses along its life cycle from mining, production, use to EoL as well as trade linkage with other nations.

(1) Case study: refining China's tungsten dominance with dynamic material cycle analysis

Tungsten is identified the critical raw material by many countries, because of its irreplaceable use in industrial and military applications. Majority of concern has been drawn to China's high share in global tungsten raw materials supply, as China holds 58% of the world's tungsten reserve and supplies around 85% of the world's tungsten. However, does China really dominate the global tungsten market? To explore the role of China in the global tungsten supply chain, it is crucial to trace China's tungsten stocks and flows from the perspective of material life cycle.

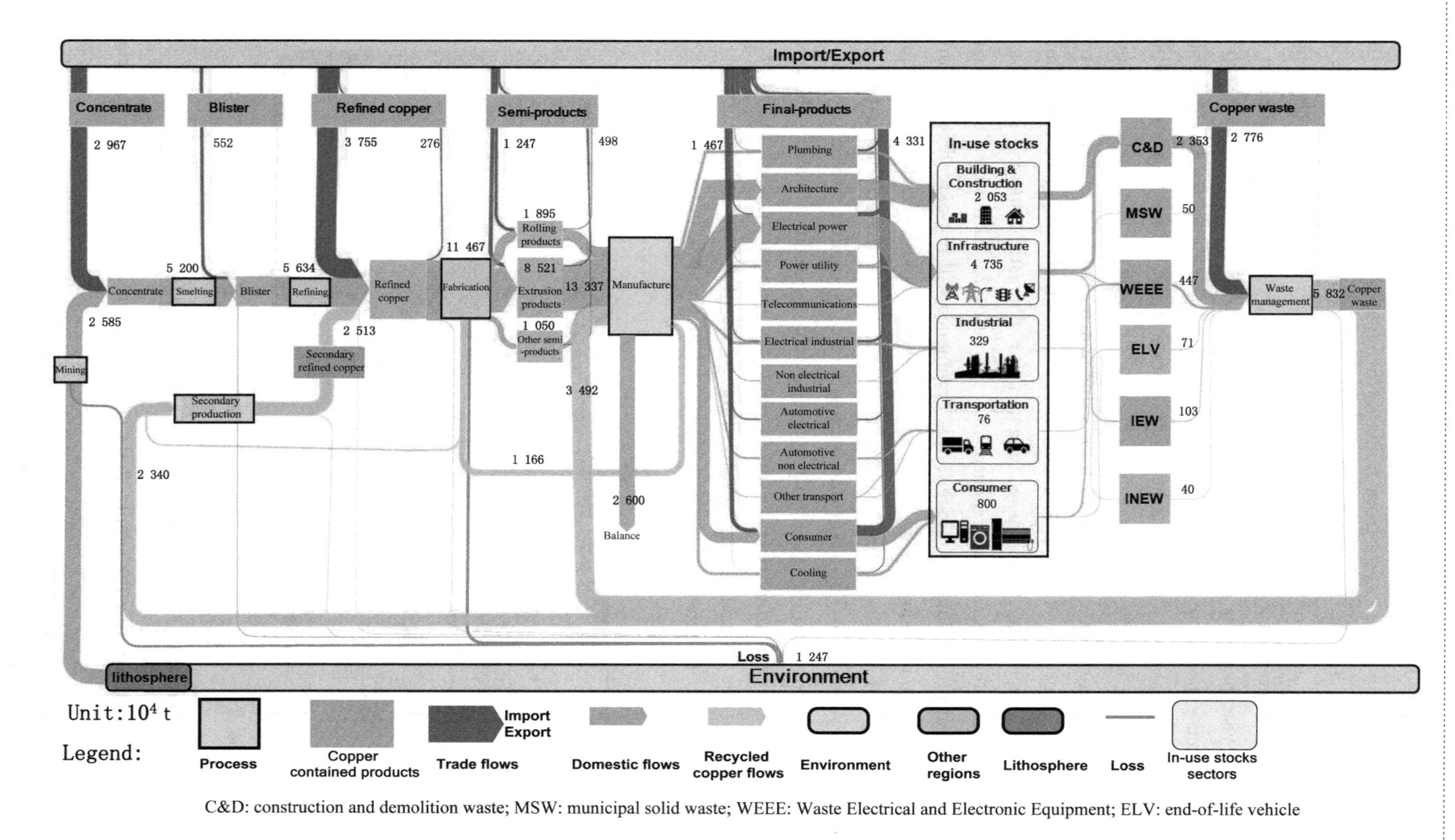

C&D: construction and demolition waste; MSW: municipal solid waste; WEEE: Waste Electrical and Electronic Equipment; ELV: end-of-life vehicle

Figure 3–2 Historical cumulative copper stocks and flows from 1950 to 2015

This case applies dynamic MFA method to trace the tungsten stocks and flows in China from 1949 to 2017 along its life cycle from mining, through to production, manufacturing, use, and EoL (Figure 3 - 3). It revealed China's contribution to each stage of the global tungsten supply chain, and provide a detailed discussion of our findings on the challenges to China's tungsten industry. It was estimated that total tungsten mined from ores in China over the past 68-year period is about 2 500 kilotons (kt). Among those, about 750 kt of tungsten have been exported to other countries, and about 970 kt tungsten is domestically consumed. It is noted about 1 720 kt has been lost from mining, production, and end-of-life stage, and merely 130 kt have been recycled as end-of-life scrap. The trade flow analysis reveals that China imported about 35 kt of high value-added downstream tungsten products from outside manufacturers, whose mineral resource was originally imported from China. It is noted that the percentage of Chinese tungsten for domestic consumption has been increasing in the past few years, and reached 58. 7 kt in 2017. Although China currently dominates the global production of tungsten, this dominance will not extend too far into the future given China's limited share of world tungsten reserves and its declining ore quality. Besides, China by itself is experiencing overcapacity issues in the primary production, which discourages the recycling of at EoL stage and makes the EoL recycling rate only about 10%. The results highlight the need for systematic measures from stakeholders along the tungsten cycle to promote sustainable practices for efficient tungsten production, use, and recycling in China. Meanwhile, this case also suggests the importance of monitoring the criticality of tungsten and other critical minerals from a dynamic and material cycle perspective.

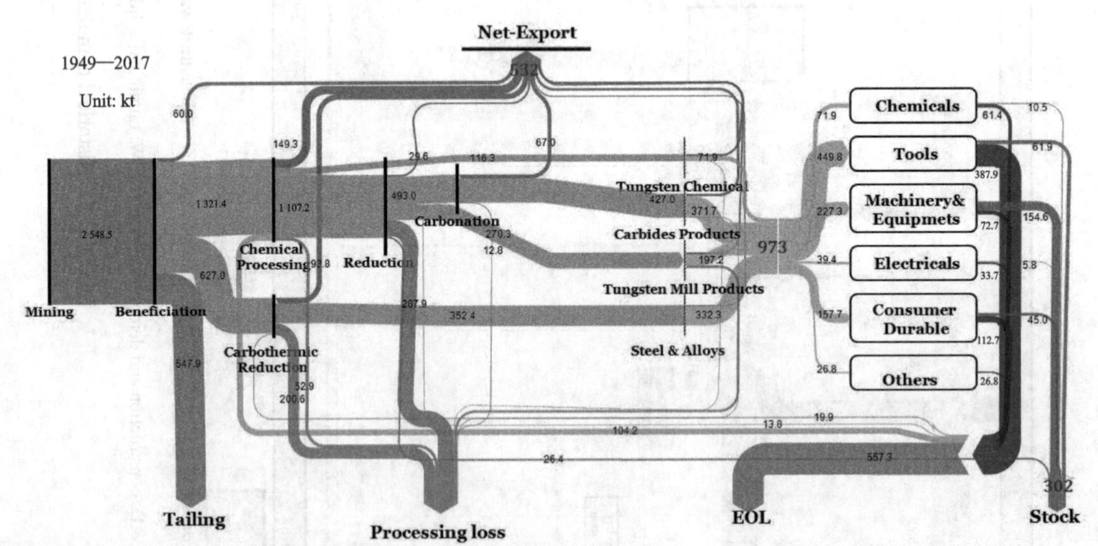

Figure 3 - 3 China's tungsten stocks and flows (1949—2017)

(2) Case study: illustrating the supply chain of dysprosium in china through material flow analysis

Dysprosium (Dy) is a heavy rare earth element (HREE) with significant uses in green technologies, which is defined as the critical materials by many countries. With increasing consumption and production of Dy, many questions arise about Dy's sustainability and environmental impacts. Thus, supply chain management is a powerful tool to solving these problems, especially for by-product metals with more vulnerable supply chains. China acts the leading role in Dy reserves, production, and exports. Moreover, Dy resources are primarily found in ionic clays and mined in south China. Therefore, a holistic and dynamic analysis of the China's Dy cycle is helpful for the global Dy supply chain management.

Thus, this work conducts an MFA framework, trace the flows and stocks of Dy in mainland China from 1990 to 2019 (Figure 3 - 4), and try to contribute to the optimization of Dy supply chains. The key findings are as follows: ① Domestic mining was the dominant source of Dy for China, with total Dy extraction being about 49 800 t over 1990—2019 period. However, nearly 50% (≈25 000 t) of total extraction was lost in the Mining & Beneficiation stage; ② The domestic Dy demand in China increased by 16-fold in the past 15 years and reached 1 400 t/yr in 2019, mainly in energy, transport, and household appliances sectors; ③ Dy mine production failed to grow in proportion to its increasing demand under intensified environmental regulations. ④ The total Dy export from China increased by 43-fold from 1990 to 2019 and had significant changes in the forms of traded commodities; ⑤ In-use Dy stocks grew by 15-fold during 2006—2019, implicating big potentials of urban mining, but commercial recycling systems have not been established. This study reveals the importance of supply-demand monitoring, environmental governance, and global cooperation to Dy industries, and highlights the necessity of material flow analysis for improving metal supply chain management, such as slowing Dy demand growth by increasing material efficiency, internalizing environmental externalities associated with Dy production, increasing the global Dy exploitation and trade, and subsidizing end-of-life Dy recycling.

3.1.2 Material flow analysis for plastics

Plastics are significant polymer compounds that are created from monomers and have greatly improved convenience in modern society. A growing production and consumption of plastic may increase the risk of leakage from plastics into the environment when they reach EoL. On the global scale, the rate of plastics recycling still at a relatively low level with only 9% of plastic waste has been recycled in 2015. Thus, plastic waste is currently one of the most troubling global environmental concerns. To address the plastic waste issue, it is important to understand plastic flows, identifying areas of inefficiency, material losses, and leakage into natural systems.

(1) Case study: dynamic stocks and flows analysis of Bisphenol A (BPA) in China: 2000—2014

Bisphenol A (BPA), like many other synthetic organic chemicals, is becoming an important part of the material basis of the modern society. Currently, the global BPA consumption has surpassed seven million tons per year, and is expected to continue to increase due to the wide use of BPA in electronics, automobiles, water bottles, and food packaging. Due to the wide use and thereby widely distributed sources, BPA leakage is

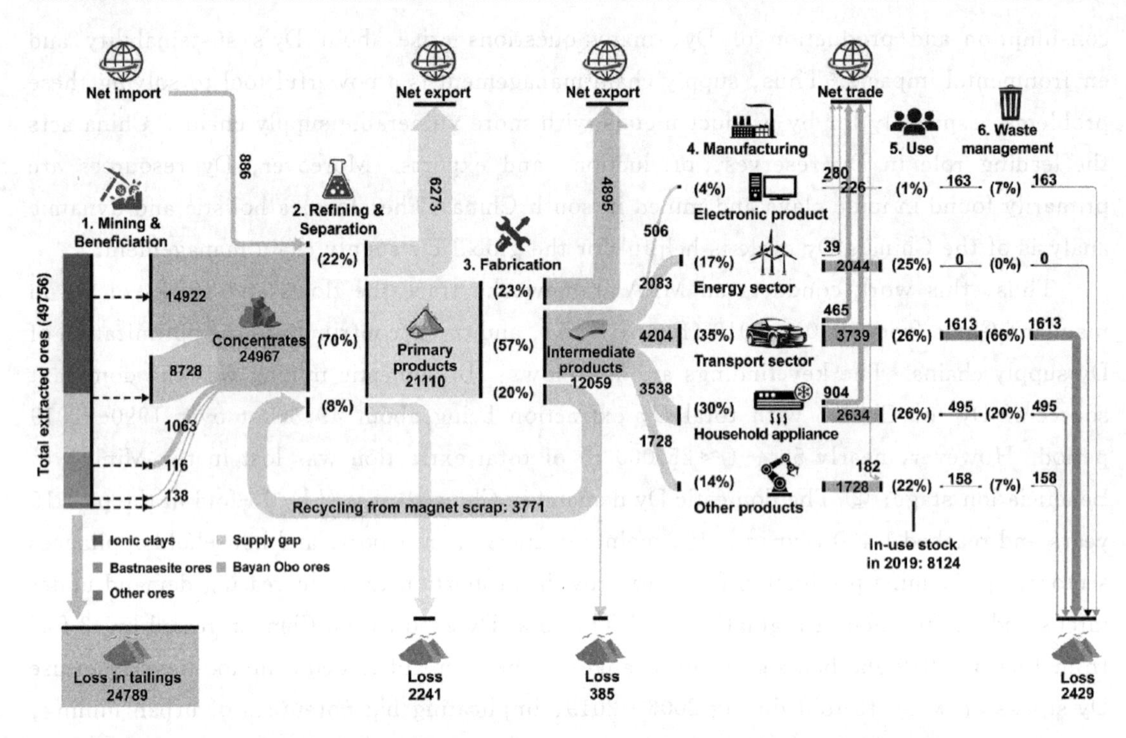

Figure 3 - 4 Cumulative Dy cycle in China from 1990 to 2019

widespread. Effectively managing the use and leakage of BPA can benefit from an understanding of the anthropogenic BPA cycles (i. e. the size of BPA flows and stocks). This case study provides a dynamic analysis of the anthropogenic BPA cycles in China for 2000—2014 (Figure 3 - 5), identifies hotspots of BPA flows on the national and regional scales, and provides a basis for understanding how BPA flows might evolve in the future.

China's BPA consumption has increased 10-fold in 2004—2014 (from 0. 9 to 2. 6 Mts). With the increasing in overall BPA consumption, China's in-use BPA stock has increased continuously, reached 14. 0 Mts in 2014. There is no clear evidence that a saturation point has been reached, but China's BPA in-use stock has increased by 0. 8 kg/capita annually from 2004 to 2014. Besides, Electronic products are the most important contributor, accounting for roughly one-third of China's in-use BPA stock. Meanwhile, the EoL BPA flows risen substantially in 2004—2014 (from 0. 12 to 0. 94 Mts). Optical media (DVDs, VCDs, CDs) make up the majority of China's current EoL BPA flow, totaling 0. 9 Mts annually. However, the flows of e-waste increase quickly, and will soon become the largest EoL BPA flow. Macroscopic BPA management strategies may need to be adjusted due to the changing quantities and sources of EoL BPA flows.

(2) Case study: material flow analysis of China's five commodity plastics

Polyethylene (PE), polypropylene (PP), polyvinyl chloride (PVC), polystyrene (PS) and acrylonitrile-butadiene-styrene (ABS) are the main five commodity plastic materials, accounting for almost 70% of total plastic demand. These five commodities of plastic have become essential to daily life for a variety of applications, including packaging, transportation, agriculture, building and construction (B&C). As the world's largest

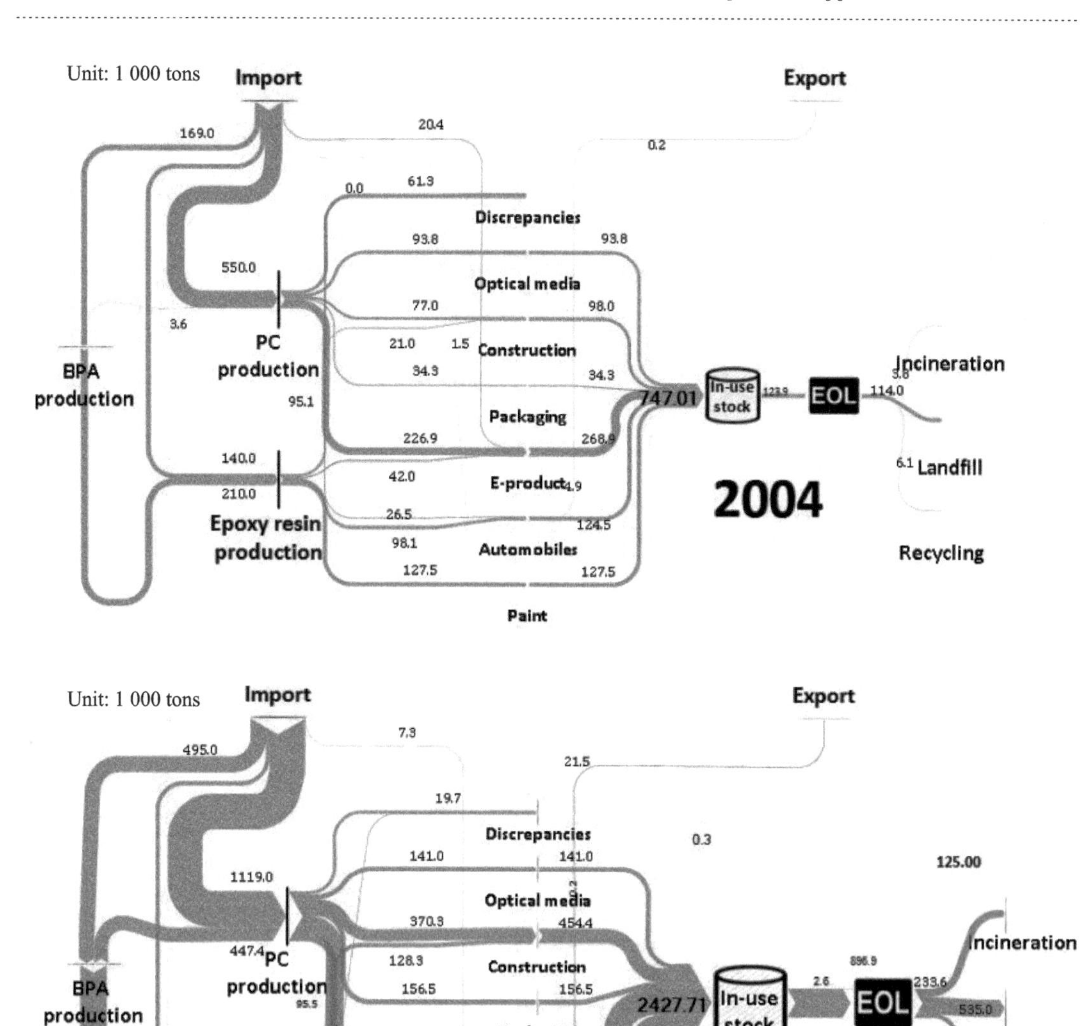

Figure 3 - 5 China's BPA flows in 2004 and 2014

producer and consumer, China is considered one of the largest contributors to the final disposal of plastic waste which causes serious environmental pollution. Thus, China has implemented various policies to promote sustainable plastic waste management. However, plastic recycling performance of specific plastic types in different applications has not been well explored in China due to its data are quite limited.

It aims to build up a unified framework to investigate the recycling status of the different applications of five commodity plastics (PE, PP, PVC, PS, ABS) in China from 2000 to 2019 (Figure 3 - 6), estimate the recycling potentials of the plastics by adjusting recycling strategies, and improve the current waste management system according to the

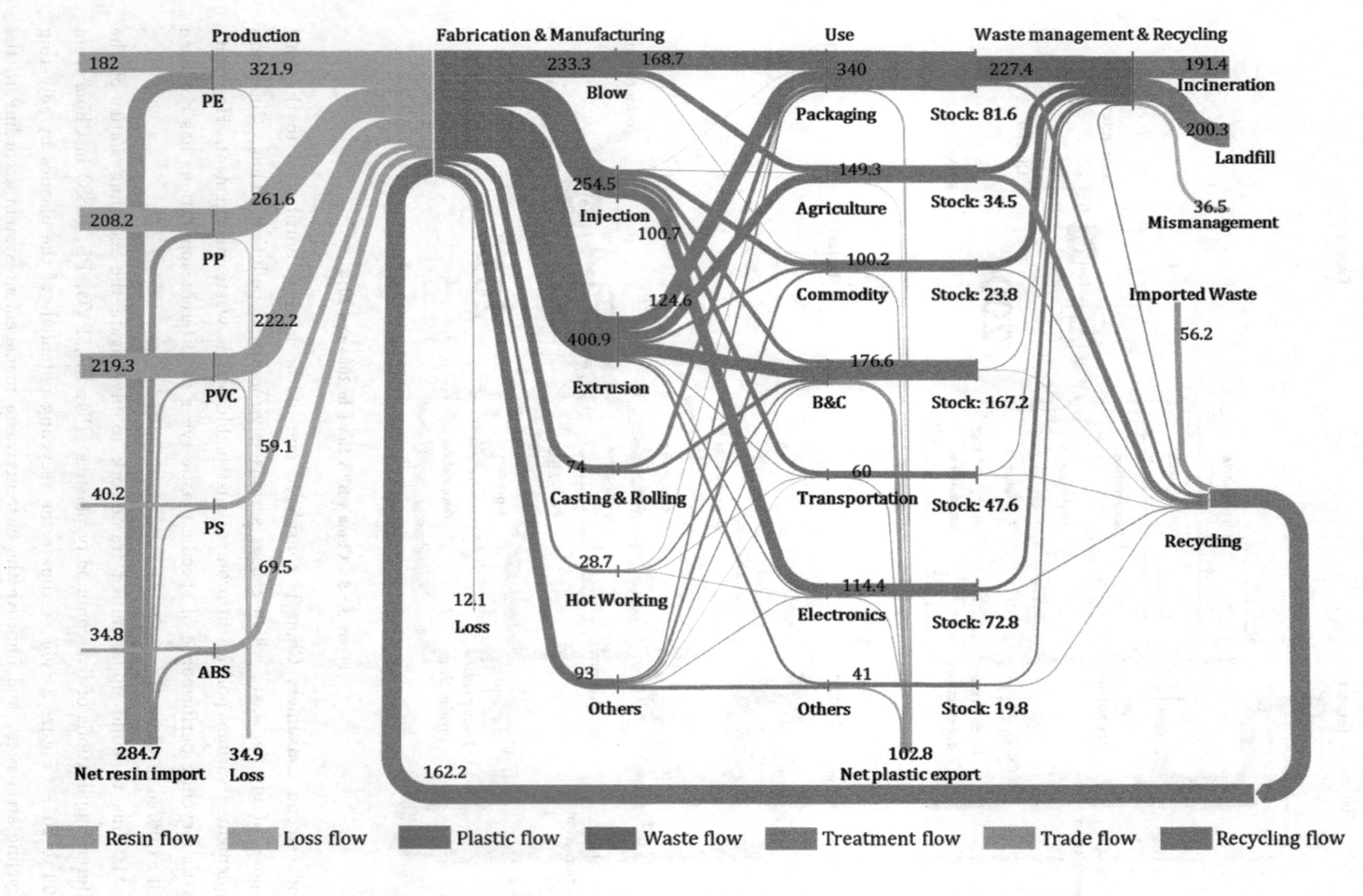

Figure 3-6 China's five commodity plastics material flow analysis from 2000 to 2019

results of quantitative analysis. The polymers amount produced from China and from other countries are around 684.6 Mt and 284.7 Mt, with 4% of them (34.9 Mt) lost in production process. In total, China consumed around 981.4 million tons (Mt) of the five commodity plastics, generating 590.4 Mt of plastic waste, only 27% of which was recycled, 34% was landfilled, and 32% was incinerated; PP (~30%) and PE (~28%) have the highest recycling rate in China, which is related to their huge consumption base, while PS and ABS have the lowest recycling rate at only ~26%; The waste recycling performance is determined by its applications, and the worst recycling rates (<20%) are the packaging and commodity sectors due to their poor collection, while higher recycling rates (≥30%) are found in the building and construction (B&C), agriculture, and transportation sectors due to the special waste collection systems in these sectors. The further examination of the recycling potential reveals that around 56% of packaging waste can be recycled by adjusting waste management infrastructure (in the collection, pre- and end-processing process). This case can help fix information gaps and support policymaking to improve sustainable plastic waste management.

(3) Case study: dynamic stock and flows of China's PET bottles

China has been the world's largest post-consumer plastics importer since 21st century. The import ban on all kinds of post-consumer plastics from 2018 issued by Chinese government significantly reshape the global structure of post-consumer plastics flows. Currently, the concrete environmental benefits of imported waste plastics have not been explored yet. Polyethylene terephthalate (PET), as one of the most widely used plastic materials (i.e. 18.8 Mt produced in 2015, about 5% of total plastics production), has become the most favorable packaging material world-wide for water and soft drinks. China has become the largest PET bottle consumer in the world since 2010, and the largest share of post-consumer plastics imported by China before 2018 is PET bottle. However, on the national level, the contributions of waste PET bottles trade on China's domestic PET bottles cycle and its corresponding environmental performance have not been studied yet, which hinders the comprehensive analysis of solid waste trade policies.

This case combines MFA and LCA to develop a retrospective analysis of all PET bottles produced, consumed, imported, and recycled within China annually from 2000 to 2018 (Figure 3-7), then provide an estimate of the environmental impacts expressed as impacts on human health, ecosystem quality, resources, and global warming associated with the handling and management of the PET material flows in China.

The cumulative recycling of PET bottles amounted to 78 Mt in China during the studied period. Among them, 29 Mt waste PET bottles (37% of total recycling) were imported from abroad and accounted for 40% of the world's total export. Most waste PET bottles in China were recycled to produce PET fibers, which significantly improved global PET circularity, reduced the use of virgin PET material, and saved about 109 Mt oil-eq of fossil resources (e.g. coal and oil) use and avoided 233 Mt CO_2-eq emissions. Despite these benefits, the environmental burdens with regional impacts during waste plastics treatment should be significantly reduced, and technologies for close-loop, namely bottle-to-bottle, recycling of PET should be further developed and widely applied.

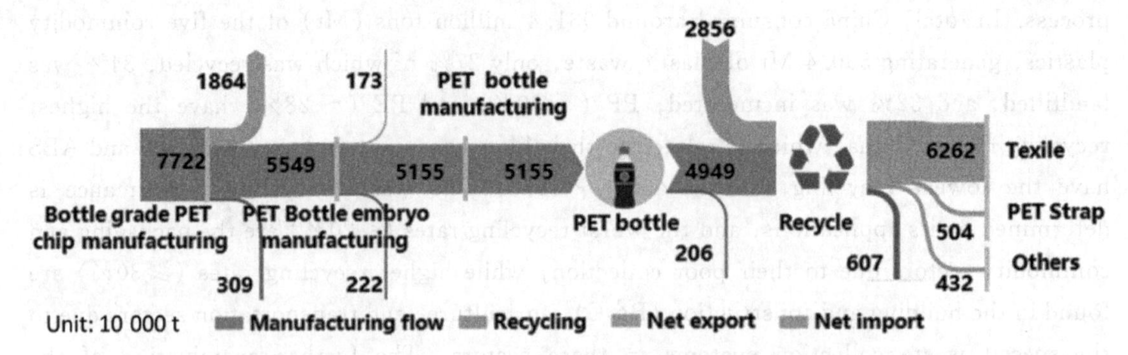

Figure 3 - 7 Material flow analysis of PET bottles in China from 2000 to 2018

3.2 Environmental management for pollution controlling

MFA has been also employed in environmental management, which is much imposed by toxic substance or organic nutrient. Phosphorus (P) is an essential nutrient for living systems with emerging sustainability challenges related to aquatic eutrophication. The biogeochemical cycle of P has been massively altered in China, challenging its food security, and causing eutrophication of freshwaters. Liu, Yuan, and their colleagues measured how P cycling in China was intensified in the past four centuries to feed up the increasing population and its demand for animal protein.

Among all countries, China, with its rapid increase in population and affluence, faces perhaps the greatest sustainability challenges in its P sector. Although China was the world's largest producer of phosphate rock, contributing 48% of total production in 2013, concerns about scarcity of Chinese P reserves have begun to emerge. Although this possibility has not been as widely recognized for P as for other critical mineral resources, reliance of Chinese agriculture on imported P would have major ramifications for global fertilizer markets. Reflecting this amplified P use, most Chinese freshwaters ever experienced excessive total P loading for years, triggering high-profile events, such as the cyanobacteria bloom in Lake Taihu in 2007 that cut off the drinking water supply for two million people in the city of Wuxi for more than one week. These pressing sustainability issues, as well as China's major role in the global P economy, highlight the importance of quantifying P pathways across China and assessing their regional impacts.

3.2.1 Anthropogenic phosphorus runoff

The model advances beyond previous socioeconomic P balances for China by including both major natural processes and anthropogenic metabolic activities with more detailed flow patterns over century time scales. Based on the assumption that anthropogenic P runoff is dependent on the intensity of relevant human activities, we further analyze the geographic patterns of anthropogenic P losses to mainland freshwaters with a high spatial resolution of 5 arc-minutes for the year 2012. We use various geographically explicit activity datasets to determine the grid-based disaggregation factors, namely, the proportion of individual human activities (e. g. cultivation) in each grid cell relative to the aggregated national totals.

Grid-based eutrophication potential factors (EPFs) are determined to evaluate the freshwater EPs induced by anthropogenic P runoff. The resultant environmentally relevant impact potential indicators, rather than estimations of actual eutrophication effects, provide more meaning to the P flow analysis results, which show only physical weights of environmental flows. An EPF is calculated by multiplying the spatially explicit fate factor (FF), describing P pathways through environmental media, with the effect factor (EF), representing the effects of a marginal P increase on freshwater ecosystems.

3.2.2 MFA of phosphorus in China

We find that P cycling in mainland China intensified from rather simple, nature-dominated stable situations in the earlier three centuries to complex, human-dominated scenarios in the past six decades, especially after China's economic "reform and opening-up" policy was adopted in the late 1970s (Figure 3 - 8). Changes of natural P flows over the past four centuries are multifarious. Natural P flow through atmospheric deposition was ~0. 3 teragrams Mt-P/y before the 1950s, but dramatically increased fourfold during the past half-

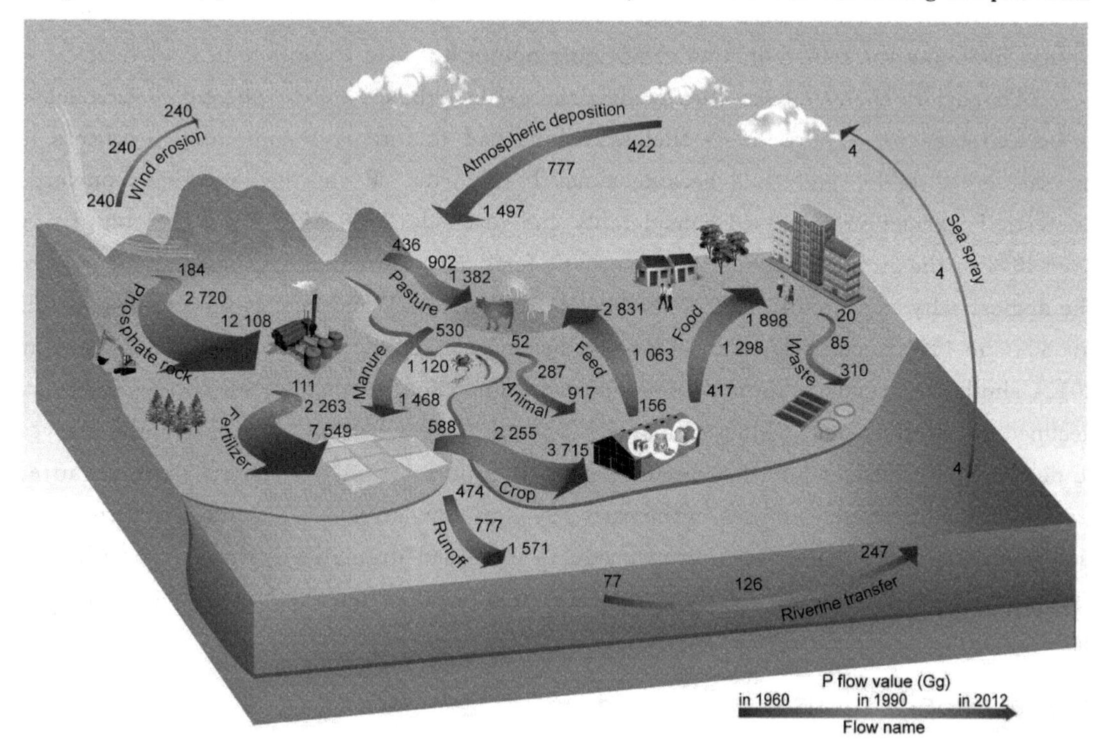

Figure 3 - 8 Schematic model of P cycles in China during 1960—2012[①]

① The thickness of the arrows denotes the historical intensity of key P flows in 1960, 1990, and 2012.

century, larger than previous estimates because of higher experimental bulk P deposition rates. Atmospheric P from wind erosion and sea spray had no significant changes and was about 40% lower than atmospheric P from combustion sources. Natural P runoff into inland freshwaters decreased from 0. 35 Mt-P/y in 1600 to 0. 25 Mt-P/y in 2012. We also found less than 20% of riverine P was transported to the open oceans annually. P entering the oceanic cycle was largely in the dissolved phase because most of the particulate P, which accounted for more than 75% of the total riverine P, was retained in the inland and coastal waters via sedimentation due to relatively flat terrains as well as artificial dams.

Before the 20th century, human mobilization of P in China was enhanced slowly through expanding agricultural activities associated with the growing population. For example, from 1600 to 1900, P contained in crops increased from 0. 44 Mt-P/y to 0. 81 Mt-P/y, and P contained in animals increased from 4. 7 kt-P/y to 13. 3 kt-P/y. However, these two P flows both decreased during the first half of the 17th century and the second half of 19th century. These declines were largely due to agricultural yield reductions when China suffered heavy population losses from decades of turmoil and wars during dynasty change and the later Taiping Rebellion (peasant revolts against the feudal monarchy). Furthermore, China isolated itself, and little P-associated international trade occurred until the first Opium War broke out in 1840. Following its defeat by the British, China was forced in the Treaty of Nanking to open four additional domestic port cities for foreign trade alongside Guangzhou. This expansion of trade led to a large-scale increase in the import of various goods into China, including the reintroduction of not only opium but also P-containing foodstuff.

This pattern changed dramatically after the 1910s, when domestic phosphate rock began to be extracted for export. Nevertheless, China had to import P-containing fertilizers to increase crop yields, and thus became a net P importer. From the late 1950s onward, domestic P extraction increased considerably and reached 12. 50 Mt-P in 2012, comprising over 40% of the global P production. Synthetic fertilizer production consumed about 70% of the domestically exploited P in 2012, followed by elemental P and feed additives, responsible for 8% and 4%, respectively. Annual P intake by crops surged to the recent peak of 3. 31 Mt-P in 2012, similar to trends in animal products and human food demands; all three, however, declined during the Great Chinese Famine (1959—1962), when millions of Chinese starved to death. Since 2003, China has become a net P exporter, and the export pattern shifted from high-quality but cheap rock in the early 1990s to downstream value-added fertilizers and fine chemicals. Net P import in crop products increased dramatically from 0. 04 Mt-P in 1990 to 0. 46 Mt-P in 2012, mainly in the form of soybeans from the United States, Brazil, and Argentina.

3.2.3 Implications

China's industrial policy has played an important role in the entire P supply chain, especially after 1949, when the People's Republic of China was established, thus implementing a centralized system. In 1983, the government raised the prices of several forms of P fertilizer, but not in proportion to their content of P_2O_5. P-associated industries were stimulated to produce low-quality but more profitable types of P fertilizers. However, on the demand side, farmers had to pay high prices for relatively ineffective fertilizer, and thus reduced their fertilizer use. This condition led to an oversupply of P fertilizers in the mid-1980s. To cope with this situation, companies had to reduce their P fertilizer production, resulting in sudden decreases in P production. Recent fluctuations in net P export may be ascribed to the changeable policies (particularly export quotas and high tariffs) imposed on phosphate rock and fertilizers to restrict P export. For instance, there was a sharp drop of net P export after the introduction in seasonal export tariffs for fertilizers in 2007.

P use in Chinese crop production has also changed dramatically during the most recent seven decades. Although Chinese farming has achieved yields close to the highest attainable for most crops, P overapplication has continued. Crop production maintained an average input intensity of 80 kg-P/ha in 2012, more than double the input intensity that crops can generally assimilate. The situation is similar in other Asian countries but different from most European countries. In addition, unlike developed countries, which are starting to recycle P-containing wastes, including composts and sewage sludge, emerging economies like China rely heavily on conventional commercial fertilizer P (over 60% of the total P inputs in 2012) to sustain agricultural yields.

Around 4% of natural P resources were eventually ingested by Chinese residents in 2012, slightly lower than the 5% in the United States. Although per capita food-P demand, defined here as the P directly consumed by humans and independently estimated from national statistics, did not change significantly in China, its composition has varied clearly over the past four centuries, especially as seen in a decline in the share of crop-P from 98% in the 1950s to 76% in 2012. These dynamics indicate that we expanded their demand for animal protein while still relying on plant-based P nutrition. The difference between per capita food-P demand and per capita food-P supply, defined here as the P available for human consumption derived from upstream flow balances, could be attributed to the direct food-P consumption of various processed products, such as bread, alcohol, condiments, and the concomitant P losses (indirect food-P consumption) during the processing stages of these products.

Changes in national strategic stockpiles, much more significant during earlier times when China stayed in seclusion, could broaden this gap as well. Crop products constitute similar percentages of food-P in Turkey, Zimbabwe, and Uganda, over 1.2-fold the percentages of food-P in Switzerland, the United States, and other developed countries. Because animal production has a much lower P use efficiency than crop production, the growing demand for animal-based diets could worsen China's impressive P sustainability challenges. Considering the P losses in diet preparation, per capita P intake (P actually eaten by humans) in China is 30% lower than in the United States but still exceeds the recommended dietary allowance (RDA). Possible reductions of per capita P intake in both countries to the RDA level could save around 219 kt-P nutrition in 2012, equivalent to the total P RDA of over 0.8 billion people worldwide. Increasingly large gaps in other socioeconomic factors between rural and urban areas, such as income levels and consumption habits, can also contribute to this shift.

The dynamics of total P mass balance in arable land over China's past four centuries can be broken into three stages: depletion (P input lags P output), equilibrium (P input equals P output), and accumulation (P input exceeds P output) periods. Only a small fraction of soil-accumulated P is liable to be chronically released to damage freshwater environments, whereas the rest, more than 85%, is likely immobilized as "legacy P", either as particulate inorganic P through chemical reaction and physical sorption or as organic P through biotic processes. However, under certain circumstances after entering freshwaters, some of these legacy materials may also release bioavailable P. Furthermore, accumulated P stocks, mainly in the form of phosphogypsum tailings and uncollected excreta, dispersed into nonarable land could exacerbate this environmental risk.

P losses to Chinese freshwaters increased threefold over the past four centuries. The anthropogenic contribution to the total exogenous P loading increased notably from around 20% in the 1600s to 83% in 2012. P runoff and leaching losses from arable lands were sizable, as confirmed by other studies, whereas, perhaps surprisingly, ever-expanding freshwater aquaculture became the largest constituent of P losses after 2009. Over 90% of the Chinese aquaculture feed-P remained in rearing waters as uneaten food and excretion, higher than levels observed in Finland and Norway (82% and 70%, respectively).

According to a 2012 US Geological Survey report, China had around 3 700 Mt of phosphate rock reserves. If we maintained its current production rate, these domestic P reserves would be exhausted in the next 25 years. This estimate is consistent with a P depletion model-based forecast, but much shorter than the world average longevity of several hundred years based upon a recent revision of global P reserve estimates that included an eightfold increase of estimated P reserves in Morocco. To address this emerging issue, China may opt to pursue more efficient and stage-based life cycle P management strategies under the combined action of multiple stakeholders. Doing so will help China to achieve the co-benefits of P resource conservation and eutrophication mitigation, which will sustain future food production and attain a healthier water environment. For example, China could delay exhausting its P reserves by over two decades by improving agronomic PUE to the average level of 80% in developed countries without impairing current crop yields.

Measures could also be taken, including improvement of aquaculture production, recycling of industrial and agronomic byproducts and shifts in human diets. In order to further prolong the longevity of P reserves further, exploitation of legacy P reserves in soils, although still in preliminary stages, would provide an opportunity for improving China's P availability in the forthcoming generations. China is becoming increasingly aware of these issues, especially in response to water quality degradation. In 2015, our government announced a goal to reduce the environmental impacts of agricultural development via policies that caped the use of P fertilizers in a new era of ecological civilization establishing. Regardless of what future trajectory China follows, the past four centuries of change in P cycles was a record of one of the most rapid and profound biogeochemical disruptions that may someday mark a boundary event of the Anthropocene.

3.3 Industrial manufacturing and production

Industrial manufacturing and production link the distribution and consumption along the life cycle. Here the publication of Gould and Colwill was chosen to illustrate the industrial manufacturing and production. Resource efficient manufacturing (REM) is a process that has traditionally been driven by market competition as manufacturers seek to reduce their production costs to maximize profits and increase sales. This primarily focuses on reducing labor, materials, and energy costs, while increasing production efficiencies and output. These previous drivers for REM have been largely economic, more recently, as REM has been adopted as a sustainability strategy, the drivers have expanded to include resource conservation, "doing more with less" clearly aligning with the goals of sustainable manufacturing and environmental improvement (like less waste, fewer resources). Sustainability-led REM will often place a greater focus on the use of materials, water, and energy, while labor and capital costs are usually secondary considerations. Therefore, while many of the past strategies, methods, and tools for REM, such as "Lean Manufacturing" and "Process Optimization" are still relevant and useful, they are potentially inadequate for identifying the additional improvements that will be required to meet the sustainability goals and manufacturing challenges of the future.

3.3.1 Qualitative material flow during manufacturing

Quantitative material flow is based upon the concept of mass balance—which is derived from the principle of mass conservation, i. e. that matter cannot be destroyed or created, although it may be converted or rearranged in space. In distinction to quantitative changes, which must always be balanced in terms of overall inputs and outputs, the conversion of matter can result in the creation or destruction of qualitative material properties. For example, the material input and output mass quantities for a rubber vulcanizing process must always balance; however, certain qualitative properties of the input rubber are destroyed upon the cross-linking of its constituent polymer chains. It can be seen that a key area for incorporating qualitative data is at manufacturing process level. The three types of material processes, as defined by MFA are: transformation, transport, and storage.

There is a large range of manufacturing processes, with a large variety of associated transformations that may affect different materials in different ways, both quantitatively and qualitatively. A key qualitative aspect of material transformations is the dependencies, which may exist between consecutive transformation processes; one process may not be able to proceed before another has finished. For example, filling a container before the container itself is produced. In this case, the manufacturing transformation, of raw material into physical container, must take place before the transformation of an empty to filled container can occur.

A further key qualitative aspect of certain transformations is their reversibility. Referring again to the rubber vulcanization process, this transformation can currently be described as irreversible. Reversibility is not an absolute term; however, as all transformation processes may theoretically be reversible, provided the knowledge, technology, and resources exist to carry out the process. Hence, reversibility may be described as a relative term, referring to the practical methods by which it may be carried out. For instance, a coating transformation process involving the combination of two materials may be described as reversible via a range of methods, such as the dissolution (a transformation process which alters the physical state of the coating) or mechanical scraping to remove the coating material. Qualitatively, the reversal of a coating transformation may only be applicable to one of the two materials; however, the coating material is further transformed by dissolution or scraping and is not returned to its original form.

The reversibility of transformations is particularly important when considering the characteristics and management of waste materials; the reversibility can dictate the possible options for waste flow. A fully reversible transformation may allow wasted material to be reused, following reversal of the transformation. An irreversible transformation may result in waste material that is not reusable or recyclable, with limited potential for recovery of constituent reusable materials.

It is not currently possible to model qualitative changes in material flow. To address this, a framework for material flow assessment in manufacturing (MFAM) has been developed, which incorporates both quantitative and qualitative information. This will form a comprehensive material flow accounting and assessment method to inform mechanisms which may be used to generate and assess alternative strategies for improving material efficiency in manufacturing.

3.3.2 MFA framework phases

The material flow assessment framework consists of five distinct phases as shown in Figure 3-9. These phases are largely interdependent, with phases 1~4 following a largely sequential but potentially iterative progression. In the fifth phase meanwhile (interpretation) runs in parallel with the other phases as an iterative and reflective mechanism, interpreting the results and applying suitable decision-making process.

(1) Phase 1: production system scope

MFAM begins with an explicit statement as to the aims and objectives of the study. The scope of the study must then be specified. This involves the definition of the system boundaries, products, and processes, and begins with the specification of the physical processes, equipment, lines, factories, etc. to be included in the study. In turn, the products that are manufactured wholly or partially by all or part of this system can then be

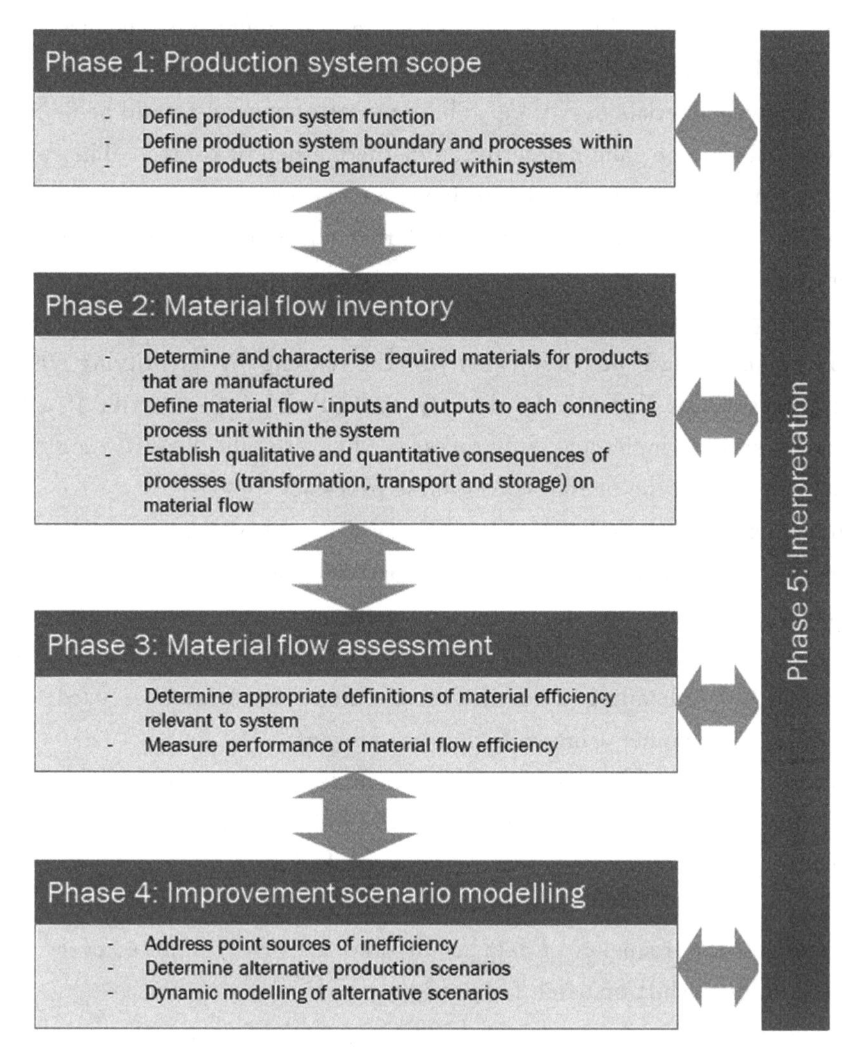

Figure 3 - 9 Five phases of the framework for MFA in manufacturing

identified, and selected or omitted for inclusion in the study. This includes defining the production system that is being examined and the products being manufactured within this system. The function and requirements of the system are clearly defined to provide a means of comparison between the performance of the existing system and any options for improvement of material flow efficiency. Thus, a subject-specific system boundary is created.

The first task in defining the manufacturing system is to state its function and purpose: what the required outputs of the system are. Both the manufactured products and the quantity of products per unit time need to be defined. Where there is flexibility in system function (e. g. where output is variable), this needs to be described. The manufacturing system must be defined in order to define the overall system boundary. The manufacturing system is defined in terms of its spatial and temporal system boundaries. This is like defining the system boundary in MFA methodology.

A system boundary should include processes and events which can be directly controlled and monitored. Manufacturing systems contain various measurement and control systems that alter the flow of materials over time. These points of control should be defined in terms of their locations, influence, and a description of the method of control. The control system may be automated, utilizing feedback mechanisms, or personnel may be responsible for (manual) process control. The automated and manual control systems, or combinations must be identified.

Manufacturing processes include transformation, transport, and storage processes. Process subdivision should be performed to the extent of identifying the smallest, fundamental process units that can be directly controlled. For example, if a controllable process contains several individual subprocesses that are not directly controllable, the boundary should be set at the overall controllable process.

Each fundamental process is connected physically with other processes, forming the various pathways for the flow of resources. Processes may be physically connected. The system boundary should describe the spatial arrangement of each manufacturing element and the physical connectivity between processes (input and output connections) including cyclic systems. This includes detailing each of the manufacturing processes used, the process grouping, zoning, or manufacturing cells that are present.

Definition of the temporal system boundary includes stating how the production system operates over time. This includes information as to production scheduling and the method of operation with respect to time for each element, e. g. batch, semi-continuous or continuous processing. The temporal system boundary should be set for each manufacturing element at a level which reflects the frequency of data acquisition, i. e. every minute, every hour, daily, weekly, or according to shift or batch frequency.

The product boundary is important to define, depending on the system, multiple products or semi-complete products may be the output. Furthermore, the inputs may be semi-complete products. This must be defined to properly understand the function of the system and how much control can be expected over the range of products that are fully or partially completed within the system. Products made entirely within the system boundary should be categorized as primary products, whereas secondary products are those which are partially processed within and partially processed outside of the system boundary.

For materials which are inputted as part of subcomponents that are pre-assembled outside of the factory, it is necessary to define if they fall into the scope for detailed investigation in this framework. For instance, the materials within the subcomponents may be within scope if the investigator can have an influence on their manufacture. If not, it is important to determine the subcomponents contribution to the material mass of the product and to assess if this is significant.

Information relating to the products manufactured should include the design and statement of the product function and service. Product groups or families contributing to the same activity (e. g. detergent, to clean) comprising specific products which vary by a particular parameter (e. g. scent and label) should be described in this detail. For example, a product group may vary by the color of a certain component; the product activity and service do not vary, but the materials vary to provide alternative colors (with the assumption that this example of esthetic variation does not alter product service). Variations in a product group that alter the product service should be described, based on the interrelation between product and service variation. Where product group variation impacts the manufacturing system, for example through the requirement for additional processes to provide the product variation, these details should be noted in the definition of the manufacturing system boundary.

Products may be grouped for assessment where variations are not considered significant. If a range of related products are made on a line within the system boundary, then the relative time spent making each variant should be quantified. The categorization of products as primary and secondary products coupled with assessing the products that are predominant within a system will allow prioritized improvement of the most significant product outputs. Once the manufacturing system boundary has been defined, it is then important to outline how material flow efficiency is to be assessed in the system. The aspects of material efficiency which are the focus should be clearly stated, to assist in the selection of criteria for assessment.

The information requirements for the study should then be stated, outlining what general information is required for the study and what information is available for the study. The data acquisition points should be determined, including location of process monitoring equipment or sensors and the frequency of data acquisition. The quality of data availability should be addressed in terms of its precision, completeness, and uncertainty. The data requirements must be set out and compared to the data availability. Any specific assumptions or requirements must be stated. Standardized processes and operating procedures may present barriers for system alteration, depending on the manufacturer and system. Similarly, any aspects of manufacturing practice that are required for certain manufacturing accreditations (e. g. GMP) must be stated and adhered to.

(2) Phase 2: material flow inventory

Inventory analysis involves attributing data and information to elements contained within the system boundary defined in the previous phase. This is achieved by defining the inputs, the processes that act upon them, and the resulting outputs. This phase creates a material flow model based on mass balance through the system, detailing both quantitative and qualitative flow within individual transformation, transport, and storage processes. The physical connectivity of process inputs and outputs is assembled to complete the system model, giving the overall product and waste outputs in qualitative and quantitative terms.

A complete inventory is required including all the substances and raw materials that are required to manufacture the products within the defined system. Material consumption during manufacturing includes both materials embedded in the final product and the materials which are not embedded in the final product but are required for production. The former can be described as embedded materials (EM) and the latter are described as auxiliary materials (AM). An example of EM is the steel embedded in the chassis of a motor vehicle and AM would be the cutting fluid and cutting tool used in producing the chassis. EM information may be collected by referring to the BOM for a product. A material may be both embedded in a product but required in the process in excess quantities. An AM may or may not be a consumable material, that is, it may be continually recycled, or consumed during the process (e. g. cutting tool erosion).

In order to create a quantitative material flow model, it is necessary to assign the quantity of each material per unit product. This provides an initial description of material requirements, which will serve as a starting point for the definition of flows through the system. This includes quantities of EM and AM; the latter should be estimated by determining the rate of consumption divided by the rate of production (in unit products per unit time). Quantitative EM information may again be collected from BOM for the product.

Material flow assessment involves investigating material flow according to a range of qualitative and quantitative measures. To facilitate this, it is necessary to include detailed semi-quantitative and qualitative material information in addition to qualitative mass flow data. The fundamental descriptors of each material should be included such as the substances which it is composed of, its primary functional parameters (physical, mechanical, esthetic, etc.), hazard information and storage information. Other inherent descriptive factors relating to waste management and environmental impact should be stated such as waste management options, material recyclability, and options for recycling, or if a material is from recycled feedstock. This inventory process serves as a database of key qualitative and quantitative parameters for the materials flowing through the system. Thus, each material in the inventory can be characterized using a range of factors that are important to material efficiency. This characterization should in part be based on self-assessment. Different manufacturers or factories have different access to services, resources, and other supply chains; for material information contained within an assessment model to be meaningful, it must be relevant to the system that is being modeled.

It is important to assign economic value to materials as they flow though processes. Economic value of materials may decrease or increase as they are transformed in manufacturing processes. For example, per equivalent mass, the value of raw steel increases as it is cast into billet steel. Subtractive processing of billet steel (e. g. cutting) produces steel off-cuts (scrap), which have decreased value. Conversely, the required steel piece may have increased value once combined with other materials to produce a high value subcomponent. A description or estimation of material accessibility and predicted future accessibility may assist in prioritizing targets for material efficiency improvement.

As it flows through processes, hazard information for a material is important to describe, and to assess the implications of a manufacturing process on the hazards posed by a material. For example, elemental mercury is more hazardous (to health) in the vapor phase than in liquid phase and this increases further when combined with a methyl group to produce organic methyl-mercury. Environmental hazards should also be described along with health hazards. Taking a life cycle view of materials requires that their recyclability be assessed and detailed. Examples of materials that cannot be recycled are vulcanized rubber and composites. Conversely, many metals such as steel can be recycled relatively easily with no loss of material properties. Recycled material inputs should be described. Recycled steel has same properties as billet steel; however, other recycled materials may have reduced or enhanced material properties. Recycled materials may have differences in texture, appearance, or other esthetic properties.

Material footprint is a description of the resources required to produce a certain mass of the raw material. For example, in general, the extraction of target metals from ores requires significant amounts of ore to be processed. The footprint of the ore may be extended to include the other resource inputs required for metal extraction and refining, as well as the transport of finished metals. Certain materials have been used in manufacturing for many decades and generally there are numerous established options for processing these. On the other hand, novel or high-tech materials generally have less well-established options for processing, which may limit the options for alternative processing routes. A manufacturer may also have limited access to alternative processing options, due to budgetary, space, resource, or legislative limitations. Thus, a description of a material's manufacturability will illustrate the practical feasibility of alternative production scenarios.

Once the input materials for the entire system have been characterized and defined, it is then necessary to define the function of process units which act upon these materials. The inputs and outputs to each process within the previously defined system must be determined. This information must be combined with information on qualitative and quantitative changes to materials during each transport, transformation, and storage process. Thus, a quantitative and qualitative material flow model is constructed that conveys a complete description of the fate of materials within the system. Including information on the changes to material characteristics in quantitative and qualitative terms (e.g. partitioning of materials and reversibility of transformations, respectively) will produce a comprehensive material flow model for interrogation in the assessment phase.

(3) Phase 3: material flow assessment

This includes steps to examine the material flow model according to various performance measures related to material efficiency. By applying different performance assessment criteria, the material flow can be assessed based on a range of metrics that appraise the qualitative and quantitative mass flow through a system. Simple mass flow metrics such as production yield can be used alongside increasingly descriptive metrics, from proportion of waste sent to landfill, to hazardous waste production.

(4) Phase 4: improvement scenario modeling

The selection and testing of potential improvement strategies takes place in the penultimate phase. At its most basic, a paper-based study can be investigated and improvement scenarios can be designed and assessed alongside the original system model, to determine the impact and define potential benefits to material efficiency according to a range of performance metrics. Alternatively, in silico simulation modeling of strategy implementation may be used to increase the scope of data handling and potential options; information collected and organized by the framework may be entered into software tools designed to simulate the implementation of improvement strategies. A separate, modified model of the system will be created (modifying system and also the material flow inventory). This will create a new model with altered qualitative and quantitative material flow, which can then be assessed and compared according to the material flow efficiency assessment phase criteria.

(5) Phase 5: interpretation

Interpretation is a continuous and iterative process that takes place during the study and is also the final phase of material flow assessment, in which the findings from the material flow inventory and assessment are considered alongside results from improvement strategy simulations and assessments. In this phase, it is determined how best to improve material efficiency in the manufacture of a product. The findings of this interpretation take the form of detailed assessment, breaking down the material flow efficiency and material processing impacts, to inform decision-making towards improvements.

3.3.3 Case study

A system is defined with the function of producing bottled beverage, in a single-product variety. The bottling system boundary is set to include the following processes: bottle molding, filling, capping, and labeling. The current production sequence is bottle molding, filling, capping, and finally labeling. The purpose of material flow assessment is to examine the waste produced in the system in terms of quality control rejects. A measurement of the number of rejects produced at each process is carried out every 100 products, giving average percentage efficiency per process. Respectively, per unit product, bottle molding adds polyethylene terephthalate (25 g), filling adds beverage (330 g), capping adds high density polyethylene (4 g) and foamed polyethylene (1 g), and the labeling process adds polyvinyl chloride (1.5 g) and adhesive (0.5 g).

Bottle molding involves the conversion of liquid material into solid material (EM) and is defined as reversible; bottles can be remolded. Filling combines the solid bottle material with the liquid beverage contents (EM). This is a reversible process but the contents must be sent to waste whereas the bottle can be reused. Capping combines the solid bottle and cap materials (EM), sealing the bottle. This process is reversible; however, the cap sealing mechanism is damaged during reversal and the cap is wasted, the bottle may be reused. Labeling involves combining the bottle material with the label material (EM) using an adhesive (EM). This process is reversible; however, a detergent (AM) is required to remove the adhesive and the label must be wasted.

The process transformations dictate the process precedence constraints: that filling and labeling must follow molding and that capping must follow filling. The rejecting rate of each process is 1%. The purpose of this material flow assessment is to determine the quality control rejecting waste produced. The rejecting rate of each process is equal (99% efficient). Thus, for every 100 products, 4 products are rejected overall and 96 products are completed. The current production sequence is bottle molding, filling, capping, and finally labeling; hence the overall cumulative yield loss is 1 102 g.

Unlike other assessment methodologies, this framework includes an analysis phase which identifies a range of alternative configurations feasible within the limitations imposed by the programmer. For this example, the process sequence configuration is examined to provide improvements, by altering the production system but not altering the processes within the system or product itself. Based on the precedence constraints for each process, there are two other feasible sequence permutations: molding, filling, labeling, and capping; or molding, labeling, filling, and capping. The former permutation has 1 099 g of yield loss and the latter produces 771 g. Thus, by reconfiguring the process sequence, a significant difference in overall yield loss in terms of mass can be the result.

The alternative scenarios generated in phase four of the framework are largely theoretical and based on the information available from the current system. Changing a process sequence may introduce additional problems which were not present in the original system. In this bottle example, labeling the bottles before filling reduces the number of rejected filled bottles from a misplaced label. However, any spillage that occurs during the filling process could now stain the label, a problem that would not occur in the original configuration. It is therefore necessary that each alternative scenario produced by the model is reviewed by one or more of the experienced factory managers/engineers and the assumptions tested. Where weaknesses in the model are identified, these can be corrected and then phase four can be repeated. As with other assessment methodologies, this phase is intended to be an iterative process.

3.4 MFA for green consumption

The recent rapid expansion of high-tech industries, along with manufacturing innovations and consumer demand, have revolutionized societal investments in infrastructure for networking and for the rapid expansion of international commerce. However, the shortening useful life expectancy of the product, driven by rapid innovation, miniaturization and affordability, and an increasingly anthropogenic metabolism have led to a major increase in the accumulation from using consumption to product waste. Meanwhile, critical raw materials have also been sinking into technical product waste, while accelerating the depletion of natural minerals.

China is not only the world's major manufacturing power, but also one of the largest consumers and exporter of products. Nowadays China has all types of industries in UN International Standard Industrial Classification System and led in the production of 220 of 500 global industrial products. EEEs and vehicles are the most fashionable aspiration of assets in Chinese households, which are the hallmark of the Four Big Items that consumers are aspiring towards. Their rapid evolution and popularity since the 1970s have led to a dramatic rise in waste accumulation and resource consumption. The consumption of some mineral resources has witnessed multiple increases, resulting in a shortage of important strategic resources and a growth of external dependence.

3.4.1 MFA for consumption process

The selection of a particular method depends mainly on data availability and robustness. During the process of urban metabolism, products flow into the consumption, then accumulate in the built environment (stock); when reaching EoL after a certain period (lifespan), they flow out as a product waste. MFA models quantitatively describe the dynamics, magnitude and interconnection of product sales, stocks and lifespans. Along this flow, China's product waste can be sourced both from domestically consumed products and imported waste (Figure 3 - 10). Domestic product waste yield is attributed to products' manufacturing, exportation, and importation. Therefore, total weight of the product waste can be defined by

$$T(x)=D(x)+I'(x) \tag{3-1}$$

where $T(x)$ is total China's product waste by weight (t), x is the year; $D(x)$ is domestic product waste weight (t); and $I'(x)$ is imported product waste weight (t).

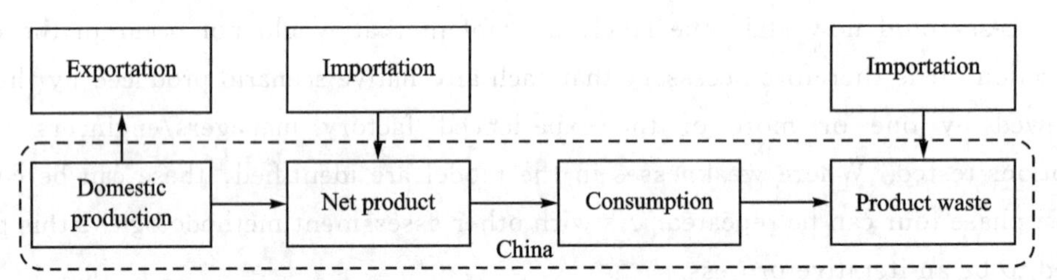

Figure 3 - 10 MFA framework for product waste generation and its boundary

3.4.2 Estimating method

To estimate China's technological product waste, we designed four steps combining: ① Data collecting and pre-mining for product consumption; ② Regression simulation for future consumption; ③ Estimating the product waste generation; ④ Validating the obtained results. Figure 3 - 11 shows the detailed research diagram of product waste estimation with the datasets and methods.

(1) Step 1: Data collecting and pre-mining for product consumption

Based on the product classification of EEE and vehicle, we collected all the available data of their production, importation, and exportation from the 1990s to 2018/2019. All the data sources are primarily from China National Bureau of Statistics (http://data.stats.gov.cn/english/) and China Customs Statistics (http://www.tradedata.hk/). The domestic consumption amount in this period can be easily determined by the equation (Domestic

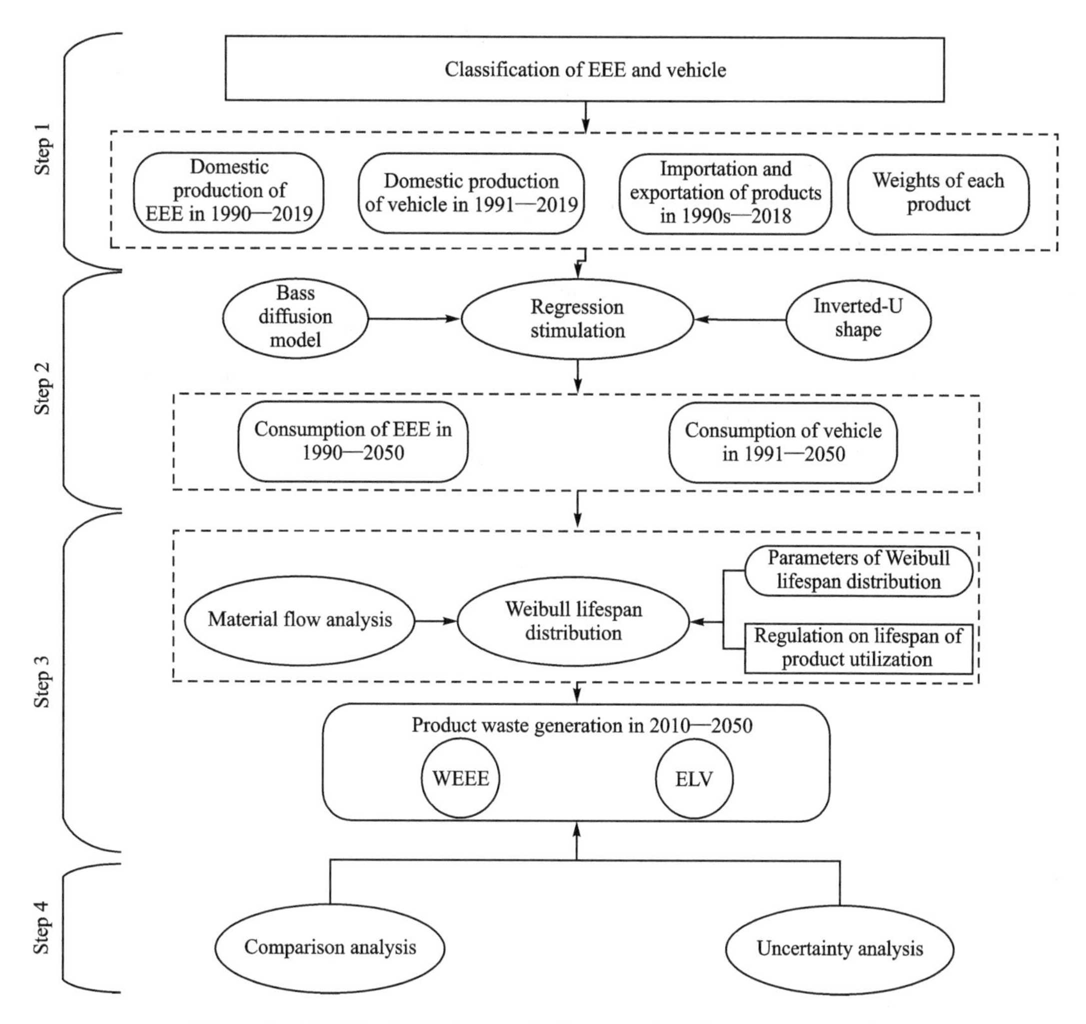

Figure 3 - 11 The detailed research diagram of product waste estimation

consumption = Domestic production + Importation - Exportation). Seventeen types of EEEs and eight types of vehicles are involved here. Some products like mobile phones are still increasing, while a couple of products such as single-machine telephone and cathode-ray tube (CRT) monitors are dropping.

(2) Step 2: Regression simulation for future consumption

In natural ecological systems, the inverted-U shape of a Kuznets curve is an adequate simulation model to address the evolution and succession of species, respectively. This theoretical model has been well transplanted into industrial ecology to forecast the future product use. But in reality, the consumption of technological products is dependent on total population, urbanization, and technology. China is projected to encounter the population peak in the 2020s and maintain a rapid urbanization in the 21st century. At the micro level, the consumption from emergence to peak can be addressed appropriately by the Bass diffusion model, which is innovatively considered for future consumption prediction. Driven by emerging technologies, the dramatic decline or obsolescence of the consumption of one product is predominately subject to the substitution of another product. For instance, the

CRT has been replaced by liquid crystal display (LCD) so that the use of CRT has been sharply cut down. Therefore, the rapid growth, stable growth, stagnation, and eventual decrease shown in the consumption model could be described approximately by exponential regression, linear regression, constant value, and linear regression, respectively.

(3) Step 3: Estimating the technological product waste outflows

Given in Chapter 2, the Weibull distribution function is adequately sophisticated to express the durable product lifespan. We use it to model the lifetime of the product. For no regulated-lifespan products like EEEs and bicycles, the probability density function (PDF) of the Weibull distribution is given by Eq. (3 - 2). Regarding the regulated-lifespan products like vehicles, the regulated lifespan should be considered and embedded into the conventional PDF for Eq. (3 - 3).

$$f(t)=\begin{cases}\dfrac{\beta}{\eta}\left(\dfrac{t}{\eta}\right)^{\beta-1}\mathrm{e}^{-(t/\eta)^{\beta}}, & t\geqslant 0\\ 0, & t<0\end{cases} \tag{3-2}$$

$$f(t)=\begin{cases}0, & t>L\\ \mathrm{e}^{-(t/\eta)^{\beta}}, & t=L\\ \dfrac{\beta}{\eta}\left(\dfrac{t}{\eta}\right)^{\beta-1}\mathrm{e}^{-(t/\eta)^{\beta}}, & 0\leqslant t<L\\ 0, & t<0\end{cases} \tag{3-3}$$

where β is the shape parameter ($\beta>0$), η is the scale parameter ($\eta>0$), and L is the maximum lifetime of products regulated in China's legislation system. EoL units for a particular duration time t can be mathematically described as ref. [28]. Based on net product production (as inflow) and the lifespan function, the annual technological product waste generation (as outflow) can be determined by Eq. (3 - 4):

$$\begin{aligned}D(x)&=\int_0^n f(x)P(x)\mathrm{d}x\\&=\sum_{i=1}^{30}[P_i(2000)\times f_i(x-2000)+P_i(2001)\times f_i(x-2001)+\cdots+P_i(x-1)\times f_i(1)]\end{aligned} \tag{3-4}$$

where x is the year; $D(x)$ is the total weight of technological product waste generation in the year x (ton); i is the ith category of product; n is the total number items in the product category; $P_i(20yy)$ is the net weight of the ith product in the year $20yy$ (ton); and $f_i(x-2000)$ is the obsolescence rate of the ith product since the year 2000.

(4) Step 4: Validating the obtained results

Two approaches are enabled to validate the obtained results. Firstly, we compare the estimated technological product waste generation to the partially real data and previous studies. Secondly, based on all the equations, the uncertainty of the technological product waste generation and their materials stock is not only caused by the variables, including each product weight and Weibull function parameters, but also on the methods like the direct linear regression simulation. Here we consider their interactive influences and adopt a MCS

(10^5 iterations) to examine the uncertainty of the obtained results.

3.4.3 Stock and waste generation

China's domestic consumption of EEE and vehicles from 2010 to 2050 is projected and demonstrated. After 2030, the total amount of EEE consumption will reach the peak of 2 500 million units. The total consumption of vehicles, however, will grow constantly during this period of time. This case also visualizes the evolution of various WEEE and ELVs outflows from 2010 to 2050. Regarding WEEE, the total unit and weight amount will grow from 635 million and 4 Mt in 2010 to approximately 4 000 million and 28 Mt in 2040, respectively. The peak time is projected to be around 2040—2045 year. With respect to ELVs, the total unit and weight will maintain a rise driven by economic growth and rapid urbanization. The total weight amount will grow from 11 Mt in 2010 to 42 Mt in 2020 and 100 Mt in 2050.

The obtained data could be applied directly in a variety of fields. For instance, the estimated product waste generation can indicate the future recycling potential of a resource: how about the economic benefit and the ascending size of the waste recycling industry? Thus, potential energy conservation and associated GHG emission reduction can be enclosed from the recycling of product waste. Last but not least, the future consumption and obsolescence of electronics and vehicles will indicate a promising consumer electronics and automobile industry in a planned way.

3.4.4 Validation

1. Comparison with existing works

Regarding five typical types of WEEE (e. g. TV, air conditioner, refrigerator, washing machine, and personal computer), their total quantity from 2010—2016 from this estimation is somewhat lower than the values reported by others previous studies. The difference can be attributed to two aspects of the previous works: old used data sources and different estimating methods. Moreover, in weight, the gaps among these studies are not significant. Regarding typical ELV quantity, some previous works indicate no distinct difference to this study. The latest data until 2019 and the scientific methods such as Bass diffusion model and Weibull lifespan distribution were employed in this study to realize an accurate estimation. Additionally, the registered vehicle data has been estimated in this study for further comparison with real value released by the government. With the variance analysis, they demonstrate almost the same value without significant difference. The difference in 2009 is much larger than one in 2016 because the available consumption data of vehicle is initialed from 1995. The data in 2009 is more sensitive to the unknown information before 1995. All the discussions can verify and validate the above results of this study, and further consolidate the relevant results.

2. Data error and uncertainty analysis

The error or range of the obtained results is primarily related to two points: one is the estimating waste methods of the Weibull distribution and the other is the projected future consumption of technological products. The error in the data of product waste outflows from 2010 to 2019 is provided from the Weibull distribution, but the error for the year of 2020 to 2050 is afforded by the two mixed points. Perhaps the projected future consumption method plays a bigger role.

We adopted the MCS to further validate the obtained results. The uncertainty of waste product generation mainly depends on each product weight (Beta distribution) and the Weibull lifespan parameters (Normal distribution). Normal distribution was assumed for the forecasted data of EEE and vehicle consumption with the standard deviation of 10%, 15%, and 20% for the data in 2020—2030, 2031—2040, and 2041—2050, respectively. We chose the WEEE weight and ELV weight in 2010, 2030, and 2050 as the examples. The increase of standard deviation indicates a growth in the uncertainty of waste product weight (Figure 3 - 12). Nevertheless, all 100% certainties can fully verify the robustness of the waste product outflows.

3.5 Waste management

Solid waste is the inevitable destination of materials derived from resource mining, manufacturing, and consumption. Its recycling significantly enhances environmental improvements and resource sustainability, which is an important part of winning the battle against pollution and promoting ecological development. Oil pollution is a global environmental problem that is related to oily sludge and contaminated sites. Due to its toxic, mutagenic, and cancer-causing properties, oily sludge, in the long run, has been regulated as hazardous waste in China, Europe, and the United States. With the development of the petroleum industry, the generation of oily sludge is gradually rising, and the annual generation rate of oily sludge in China has reached approximately 3 Mt.

Oily sludge is notably generated during the processes of petroleum exploitation, processing, storage, transportation, and generation in the oil and gas industry. China has regulated it with typical waste codes of 071-001-08 and 071-002-08 within *Hazardous Waste List 2021*. It is characterized of large generation, massive storage, and intractable disposal. Additionally, its physical and chemical features differ in their generation sources. In general, the oil content in oily sludge is between 10% and 50%, and the moisture content ranges from 40% to 90%. Therefore, if not treated properly and exposed to the environment, oily sludge is prone to cause oil leakage and heavy metal release.

3.5.1 Material flow analysis of mechanical separation treatments

Currently, the mechanical separation is widely employed to treat oily sludge in China. We use a reported operation at an oil generation plant in Heilongjiang as an example and establish a material flow diagram for this microlevel. Through original data collection and estimations of the generation, primary treatment, further treatment, comprehensive

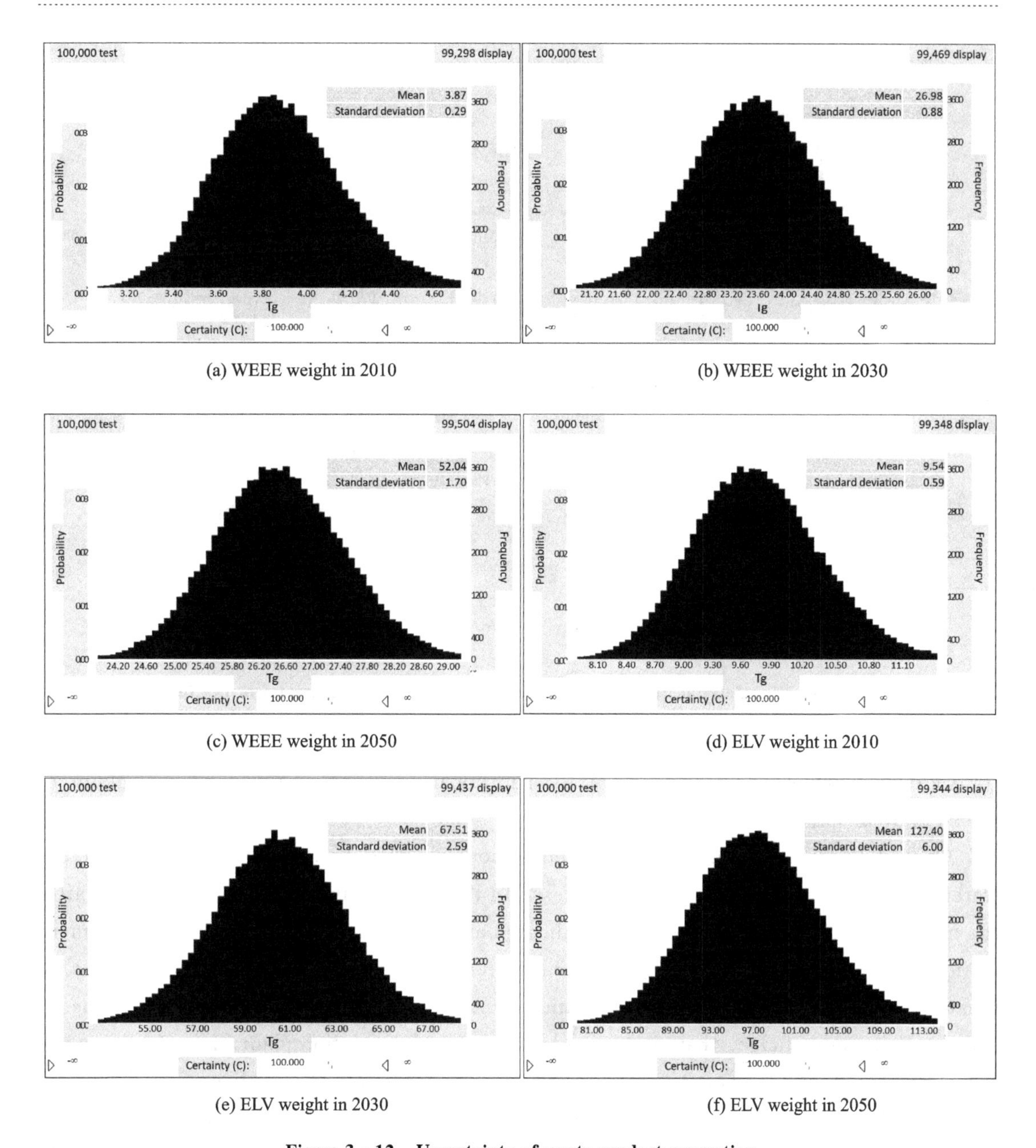

(a) WEEE weight in 2010 (b) WEEE weight in 2030

(c) WEEE weight in 2050 (d) ELV weight in 2010

(e) ELV weight in 2030 (f) ELV weight in 2050

Figure 3 - 12 Uncertainty of waste product generation

utilization (i. e. recycling and recovery), and disposal, a material flow diagram of the mechanical separation for processing oily sludge was obtained (Figure 3 - 13).

In this case, the average annual generation of oily sludge was around 41.5 kt (1 kt= 1 Gg= 10^9 g). After the treatment stage, on average, around 25% of oily sludge was disposed of in landfills. An average of 4.2 kt of substances flowed to the comprehensive utilization stage and accounted for 10% of the total annual generation. Among them, the substances entering the comprehensive utilization stage from the mechanical separation and dehydration processes are mostly recoverable crude oil with average volumes of 1.8 kt and 2.4 kt, respectively. The substances that flow from the mechanical separation and

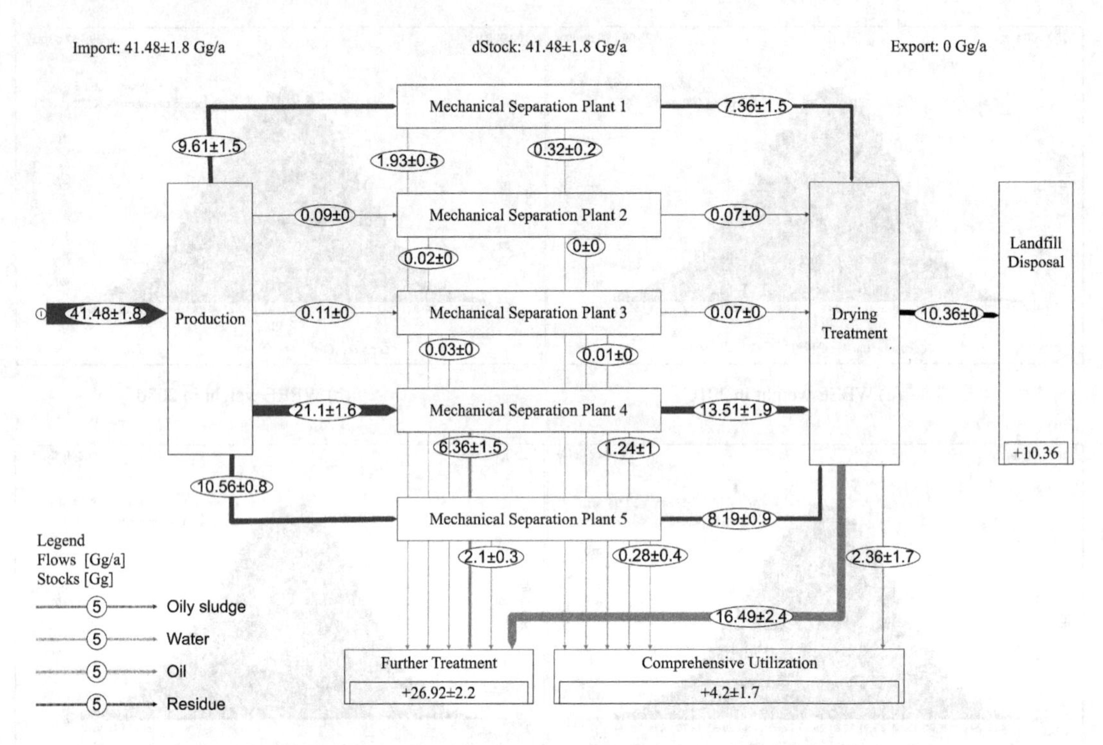

Figure 3 - 13 Material flow of oily sludge mechanical separation treatment in China

dehydration processes into the following treatment stage are mainly wastewater that requires transport to a sewage processing plant for additional treatment, with average amount of 10. 4 t and 16. 5 t, respectively.

Overall, by virtue of high moisture and oil content, the separation of crude oil from the water is the crucial core in the whole process. In this case study, the sludge remaining after the mechanical separation process is consolidated and transported to a dehydration treatment station. Since the oil content of the remaining sludge after the dehydration process may be greater than 3%, the landfill disposal ultimately needs to be adopted. Nevertheless, the oil content of the oily sludge, that was cured by mechanical separation plant 2, conforms to the requirements of the comprehensive utilization standards of Heilongjiang. Thereby, if physical and chemical component control tests are conducted on the treated sludge of each mechanical separation plant, and the treated sludge from each plant is transported separately to corresponding, different dehydration treatment stations for the next process, the comprehensive utilization of the whole process can be greatly elevated.

3.5.2 Material flow analysis of incineration treatment

The incineration process is one of the earliest oily sludge treatment methods adopted in China. By collecting all the relevant data, the material flow diagram of the treated oily sludge by the incineration treatment process was obtained (Figure 3 - 14). The generation stage represents the process of producing 1 ton of oily sludge from an oil field. The treatment stage consists of two processes; the filter pressing process, and incineration process, and it shows the amount of treated oily sludge by each process, as well as the amount of oil and water that are separated through the treatment process. The further treatment stage includes those substances that need additional processing. The comprehensive utilization stage consists of resource recovery and recycling, for example, the separated crude oil during the treatment stage and the sludge that can be comprehensively utilized. In this case, the released residue in the incineration process corresponds to the comprehensive utilization standards.

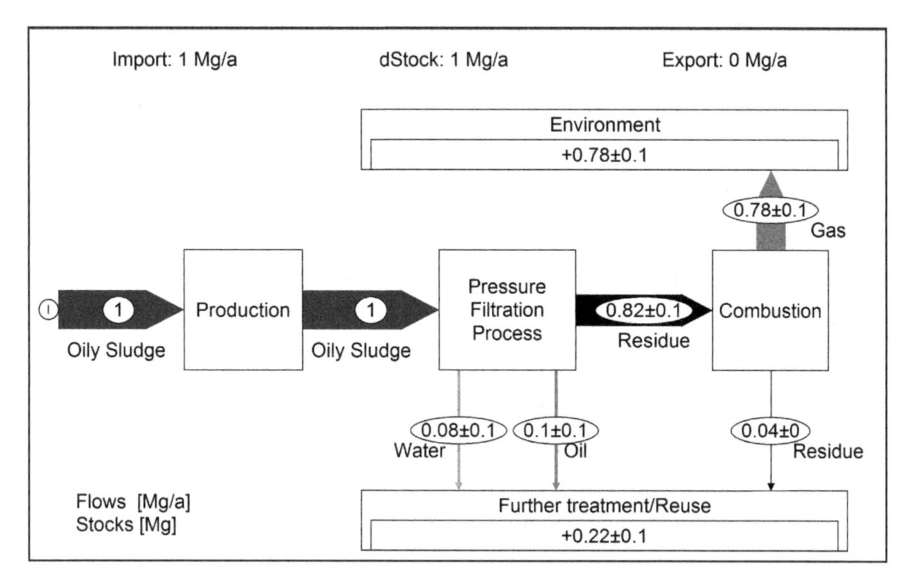

Figure 3 - 14 Material flow of the oily sludge incineration treatment

From the filter pressing process, water can be separated and transferred to the further treatment phase, and the crude oil can be transferred to the comprehensive utilization stage for recycling, with each accounting for 8% and 10% of the total amount of treated oily sludge, respectively. After the filter pressing process, on average, 82% of the treated oily sludge will move to the incineration process. During this process, 78% of the total treated oily sludge is finally combusted to produce waste gas, and was discharged into the atmosphere after reaching the standard through the spray tower, while 4% of the total oily sludge becomes ash residue. Later, by comparing the physical properties of the ash residue with the comprehensive utilization index, we conclude that, the ash residue meets the requirements of the comprehensive utilization standard. Eventually, the ash residue can be transported to the comprehensive utilization plant to be used for road paving, well site laying, waste sealing, and waste covering.

From the material flow diagram of the incineration treatment process, the comprehensive utilization could reach 14%, which covered crude oil and ash residue reuse and recycling. However, it is worth noting that in this case, the generated ash and waste gas during the incineration process meet the environmental requirements, so there are no impacts on the environment. Nonetheless, this does not mean that all the remaining residues and waste gas was innoxious. The physical properties and chemical compositions of the generated ash and waste gas depend upon the properties of the treated oily sludge, incineration temperature, ventilation, and other factors.

3.5.3 Material flow of China's whole process management of oily sludge

By running some assessments and predictions, the total average generation of oily sludge was approximately 5. 3 Mt, which was derived from 16 different oil fields. It can be seen from the flows on the left side of the diagram that the Daqing oil field had the highest annual generation of oily sludge and was followed by Dagang, Changqing, Yanan, Zhongyuan, and Henan, while Fushan had the lowest annual generation of oily sludge (Figure 3 - 15). With our field survey in Hebei, 11 companies in 2019—2020 were involved to produce 3 320 t, and around 3 307 t were tackled.

The middle part of the diagram shows the seven leading treatment processes consist of mechanical separation, pyrolysis, thermochemical cleaning, solidification, dehydration, superheated steam injection, and incineration (Figure 3 - 15). The most widely used approach was thermochemical cleaning, by which its recycled amount accounted for 42. 13% of the total generation. The second most used method was mechanical separation, by which the amount of treated oily sludge made up 25. 28% of the total generation. Dehydration, incineration, solidification, and superheated steam injection follow in importance, and the amounts of treated oily sludge by each treatment proportion were 7. 95%, 6. 84%, 3. 42%, and 2. 17% of the total annual generation, respectively. In the lower-left corner of the diagram, several oil fields are demonstrated, but the used approach for 12. 21% of the total generation was still unknown.

As seen in the treatment stage (Figure 3 - 15), the processes with the highest amounts of water separation included the thermochemical cleaning, mechanical separation, and incineration processes, with average annual dewatering amounts of 1,291. 6 Gg, 292 Gg, and 268. 5 Gg, respectively. However, the amounts of water separated from the solidification, pyrolysis and superheated steam injection processes were relatively low at 158 Gg, 126. 2 Gg, and 94. 5 Gg, respectively. Among all processes, the two procedures with the highest oil separation amounts were the thermochemical cleaning, and the mechanical separation, which annually occupied total 323. 8 Gg and 126. 9 Gg, respectively. The amounts of separated oil during the pyrolysis, dehydration incineration, solidification and superheated steam injection processes were relatively low at 34. 2 Gg, 24. 7 Gg, 24. 6 Gg, and 12. 5 Gg, respectively.

On the other hand, the treatments with the highest dehydration rates were the superheated steam injection, pyrolysis, dehydration incineration, and thermochemical cleaning treatments, which reached the levels of 81%, 69%, 63%, and 57%, respectively. Furthermore, the dehydration rates of the solidification and mechanical separation treatments were around 43% and 22%, respectively. Comparing to the superheated steam injection, mechanical separation, solidification, and dehydration incineration, pyrolysis and

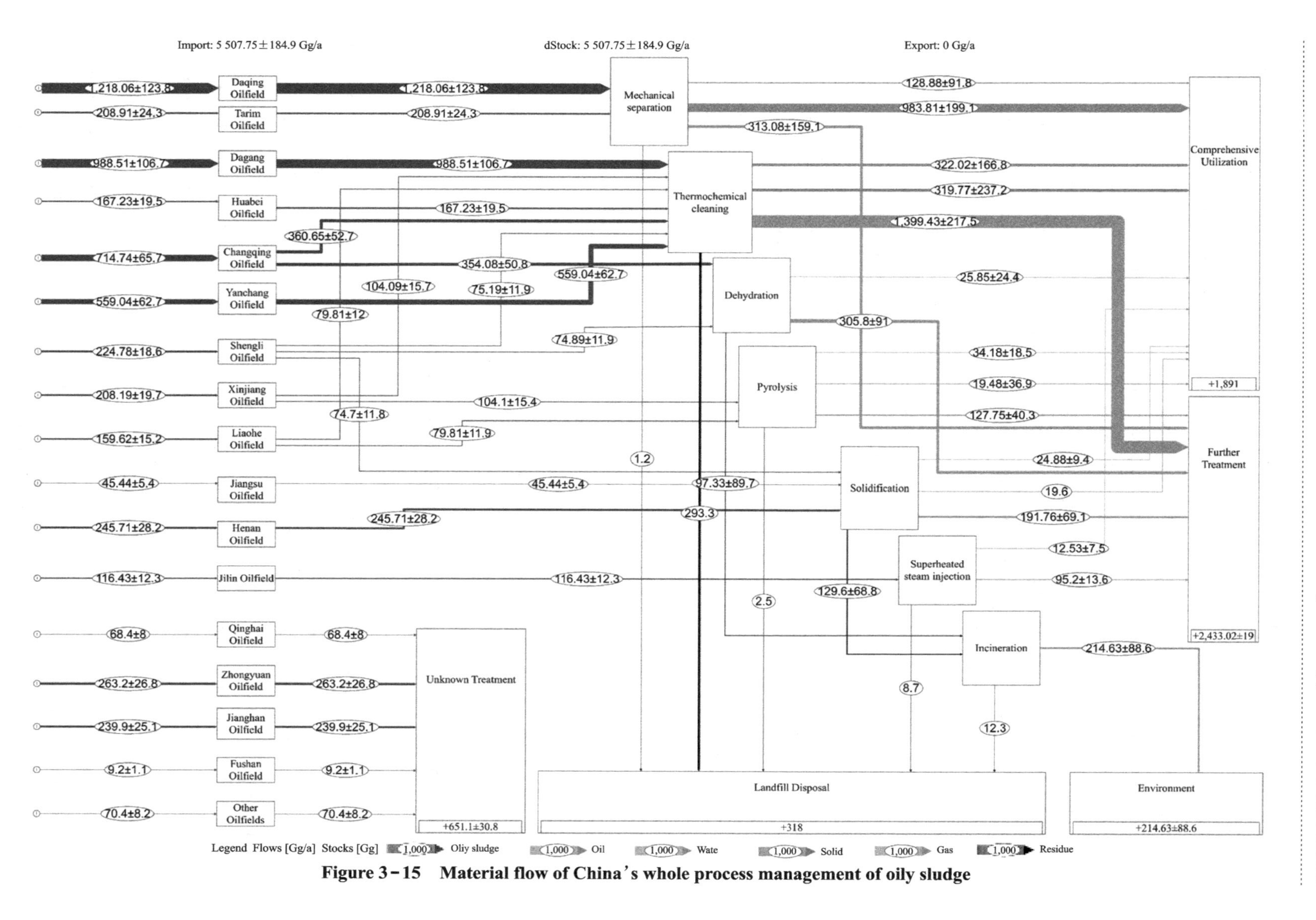

Figure 3-15 Material flow of China's whole process management of oily sludge

thermochemical cleaning treatments had extremely high oil separation rates, up to 19% and 14%, respectively.

The Daqing and Tarim oil fields primarily employed a mechanical separation, and the comprehensive utilization amount was approximately 927.5 Gg per year. The comprehensive utilization rate of mechanical separation has grown from 9.42% to 78.24%. It indicates that 68.82% offered by mechanical separation could satisfy the local standards. The Yanchang, Changqing, and Xinjiang oil fields chiefly adopted the thermochemical cleaning, and the comprehensive utilization amount after treatment was approximately 337.5 Gg. The comprehensive utilization rate has increased from 14.42% to 29.45%, 15.03% of which was imposed by thermochemical cleaning. The Xinjiang oil field also adopted pyrolysis approach so that the comprehensive utilization amount was approximately 19.6 Gg. Therefore, the comprehensive utilization rate of pyrolysis has grown from 18.75% to 29.51%. In this regard, 59.9 Gg of sludge belonged to comprehensively recycling. Solidification has played an important role in increasing the rate of recycling from 7% to 23%.

Sludge entering the incineration process was consisted of 130.7 Gg of the remaining sludge after the drying and 122.1 Gg that cannot be recycled. In the whole process, 95% of the total incineration capacity turned into waste gas, while only 0.05% became waste residue. The disposal of oily sludge was majorly destined to landfill. Part of the landfilled sludge was derived from mechanical separation, thermochemical cleaning, pyrolysis, and superheated steam injection. The disposal amounts were approximately 1.2 Gg, 293.3 Gg, 2.4 Gg, and 8.7 Gg, respectively, and the disposal rates for each process were 0.09%, 13.06%, 13.20%, and 7.5%. The other landfilled sludge was deprived from incineration after solidification and dehydration treatment, whose volumes were 6 Gg and 6.4 Gg, respectively. As a result, approximately 317.9 Gg of sludge, 6% of the total generation, flowed into landfills every year.

3.5.4 Implications for oily sludge management

Overall, the annual generation in China ranged from 4.45 to 6.22 Gg. Of this amount, about 88% were mainly treated by seven approaches. The predicted average comprehensive utilization rate was 35%, for which the rate for oil substances accounted for 10% and the rate for treated sludge occupied 25%. Different oil fields have adopted a variety of techniques, as a result, dehydration, oil recycling, and local comprehensive utilization have differed. Their standard was only implemented in specific areas. Therefore, the sludge was likely tended to a comprehensive utilization, but because adequate policy and standards are still lacking in the whole nation, part of the sludge would eventually disappear probably in an informal process or be destined for dumpling.

Recycling through a circular economy philosophy is recognized as a unique solution to environmental pollution and resource depletion. It is rapidly reforming the whole anthroposphere related to manufacturing, producing, consumption, and recycling. Some negligible toxic substances can be found in oily sludge so that its recycling is susceptible to environmental risk. Thus, circular economy for oily sludge may not always be advisable. Circular flows involving toxic substances may impose a high risk on the environment and public health, such that overemphasis on circularity is not desirable. Although new environmental laws and the Soil Pollution Control Action Plan in China have been implemented and solid waste management laws are in their second revision, close supervision and technical processes should be greatly strengthened to mitigate the oily sludge problem.

3.6 Regional metabolism and management

As the places where people live and work, the regions and cities play the fundamental role in the transition from a linear to a circular economy. Additionally, regions are not closed. They are open to the outside world with constant exchange of energy, materials, and information. The regional agendas for the circular economy call for conceptual and methodological development to characterize the material flows that underpinning the regional economic and social development.

3.6.1 Concept of regional metabolism

Natural resources provide the material base for regional social-economic system. The human society retrieve all kinds of natural resources from nature, using them in production by transforming their physical characteristics and transporting the products from one place to another. Along the flows of the materials within human society, new value is produced with the input of labor, while wastes are also produced and discharged to nature as unwanted by products. During these complex processes of production and consumption, the regional social systems and surrounding ecosystems are complex, they interrelated with each other, reproducing themselves and co-evolving over time like any living beings.

Regional metabolism applies the biological concept of metabolism to regional social-economic systems, by charactering a set of inflows and outflows of materials, energy and water at various stages of consumption, use and storage within a specific region. The scales of the region can range from local to national, and even cross national. The quantification and subsequent analysis of these flows aims at identifying synergies between the consumption needs of economic activities and the availability of materials for production, and evaluating the resources efficiency for regional development. Here, the metabolism is not used as a metaphor, because the material and energetic process within the human society is in consistence with that of the natural systems, and forms the intrinsic linkage between human society and the nature.

The concept of regional metabolism provides a systematic approach on the implementation of circular economy with the territorially based actions adapted to their specificities. Therefore, it connects to three interrelated components for the overall framework of circular economy in cities and regions (Figure 3 - 16):

① The megatrend of physical material flows within the regional society that dynamically

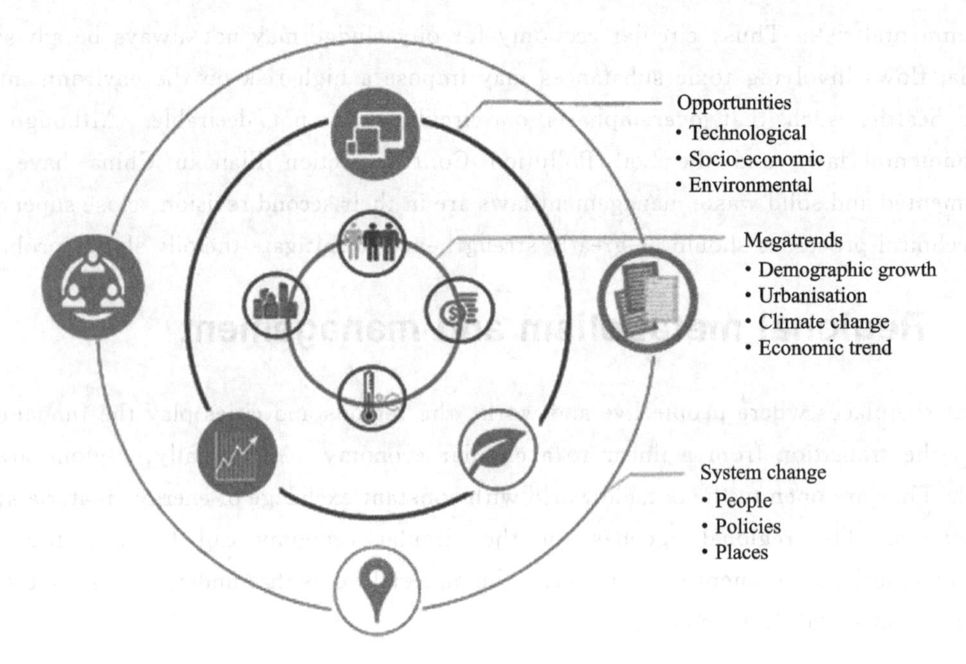

Figure 3 - 16 A conceptual framework for circular economy in cities and regions①

reflect the specificities of the territory;

② The technological, socio-economic, and environmental opportunities for effective improvement in both resource efficiencies and social wellbeing;

③ The systematic transition at the regional level, which shapes the landscape in the long term.

3.6.2 Economy-wide MFA for regional metabolism

Economy wide material flow analysis has been implemented as a guiding tool at the regional level to characterize the status and trend of regional metabolism. As part of a regional environmental management and audit system, EM-MFA can be used to identify the overall resource consumptions and waste output at various regional level, as well as its effects on resources efficiency. Thus, the calculation results are used for improvement in regional value added based on local resources endowment, and promotion of new technologies to reduce environmental pollution.

1. EW-MFA framework

In 2001, Eurostat established a framework and standard for economy wide material flow analysis, which can be used at various regional scales. These economy-wide material flow accounts and balances show the amounts of physical inputs into a region, material accumulation in the region and outputs to other regions or back to nature. If considering the cross-border effects of the flows, the scheme can also include indirect flows associated to imports and exports through the economy as illustrated by Figure 3 - 17.

① Source: OECD, 2020.

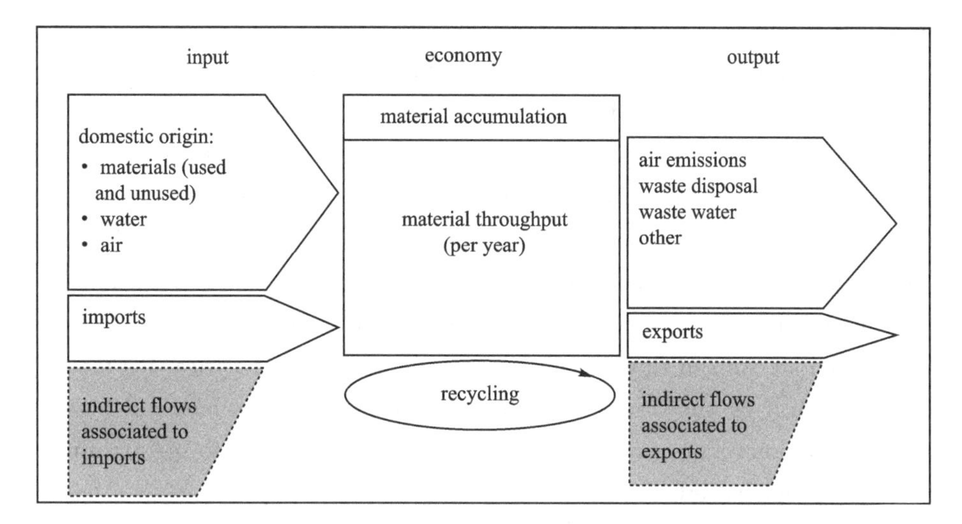

Figure 3 - 17 Economy-wide material flow accounts for a region (excluding air and water flows)[①]

2. Implementation of regional metabolism approach in regional management

Regional metabolism approaches can bridge natural and social sciences in addressing sustainability and providing useful information for monitoring and policy-making in addressing regional agendas for circular economy. The material flow analysis is the fundamental tool for regional metabolism studies, which can help to solve problems related to territory-based environmental management within a specific region, including:

(1) Recognition of the trend of regional environmental loadings

Regional socio-economic systems constitute hybrids of biophysical and symbolic systems shaped by discourses, power relations or monetary flows, and are subject to social organization and institutions. Generally, the societal resource use kept accelerating in last decades in many regions in the world due to rapidly progressing industrialization and urbanization. If the regions in the emerging economy follow the paths of the high-income countries, the growth of resource use per capita will keep growing with the rising economic activity and affluence. That is why it is important to include the indicators from regional metabolism into the index for sustainable development goals (SDGs) to better represent the environmental burdens of the regional development.

(2) Identification of key challenges at regional level

As the regional development and socio-economic structural change, the surge of resource use challenges the sustainability and safety of regions, where political competencies for resource management exist. Regional metabolism has developed various indicators that can be applied in sustainable resource use policies in different local context, while sharing a similar framework that adaptable to SDGs. The data on extraction, trade, processing and consumption of resources in different regions provides indicators from both production- and consumption-based perspectives to characterize both the demands on local resources, and the

① Source: Eurostat, 2001.

impact on global supply.

(3) Setting of policy priorities for regional circular economy

As address above, the regional metabolism study is a critical component for regional circular economy, which has gained substantial traction in many countries. It is necessary to have a detailed description of material cycles in a region before developing any sector-, material- and product-specific strategies and policies to foster circularity. The regional material flow analysis tools can help us to take a closer look at the flows within the regional socio-economic system. The material cycle perspective provides a consistent framework to depict the material stocks and flows in different regions, which is critical for identifying the policy priorities, such as show how fast material stocks grow, when and how materials become available for recycling, and how much recycling contributes to maintaining stocks.

(4) Designing regional infrastructure for closing of material cycles

The circular economy implies a systemic shift, in which services are provided with efficient use of natural resources. In order to have the economic activities being carried out in a way to close loops across value chains, regional infrastructure need to be redesigned and rebuilt to unlock the linear lock-in. The efforts by individuals to reduce pollution emissions, to save natural resources, and to increase the share of renewable resources, requires supports of public goods of regional infrastructure. Public investment is needed for the planning and construction of the regional infrastructure for circularity, which has potential for new economic growth decoupled from the growth of environmental burdens. Moreover, the material close-loop infrastructure facilitates the activities, such as repair, maintenance, upgrading, remanufacturing, reuse, recycling, which are more labor intensive than the mining and manufacturing of a linear economy. The regional circular economy is likely to provide more jobs for social inclusiveness.

3.6.3 Case studies: Regional Metabolism in Beijing

Using the framework of EW-MFA, the metabolic model of the throughput in Beijing was established to show the change of material flows in the city from 1992 to 2014, when the city experienced dramatic population growth, spatial expansion, and socio-economic transition. In 1992, Beijing had a population of 11 million, and the GDP was 71 billion yuan; while in 2014 the population reached 21.7 million, and the GDP grew to 2 292 billion yuan.

1. Implementation of EW-MFA framework

The original data for material flows of different categories was retrieved from various official statistic yearbooks. The determined results are shown in Figure 3 - 18:

2. Implications to regional policies

As the national capital of China, Beijing has experienced dramatic change in last decades. The open and reform of the country reshape the fundamental mechanism of

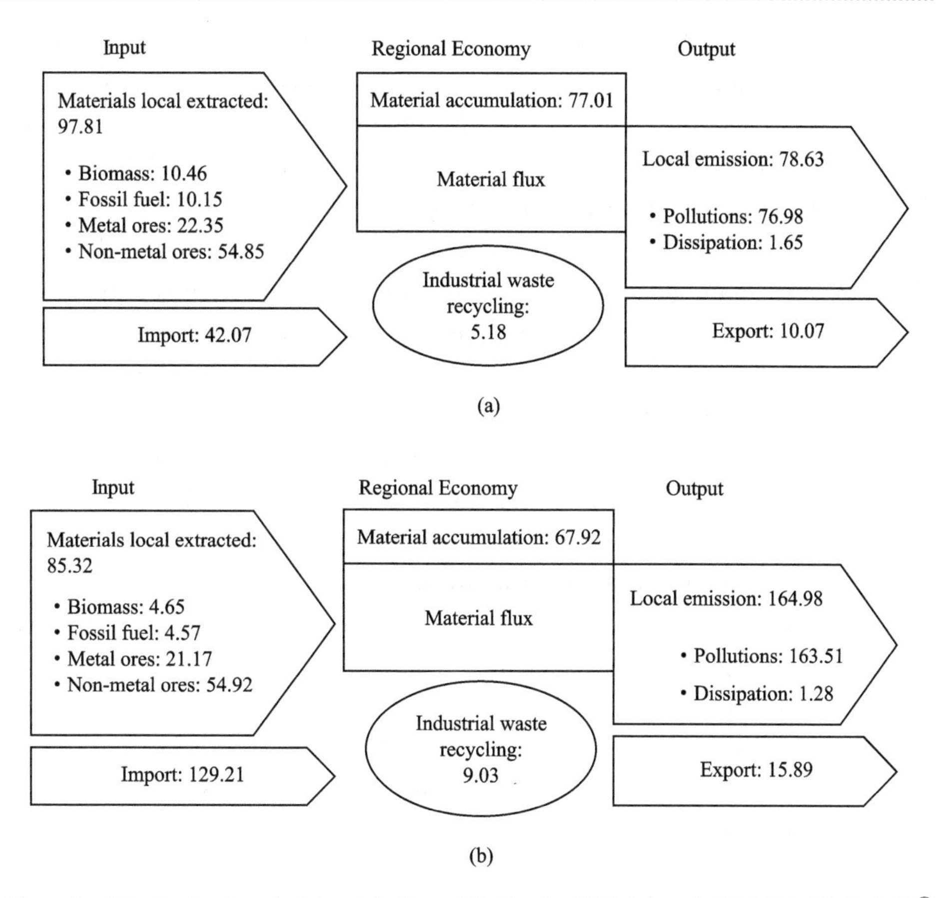

Figure 3 - 18 Regional material metabolism of Beijing in 1992 (a) and 2014 (b) (Unit: Mt)①

economic and social development, as well as the relation between urban and rural areas. Such social transition had deep impacts on the material flows of Beijing, and its surrounding areas.

(1) Trend of regional environmental loadings

The results indicate a general upward trend in material input and output of Beijing from 1992 to 2014. However, considering the population growth, the resource input per capita had decreased. Minerals and fossil fuels used to be the major parts of its material outputs before 1990s, when steel production and petrol chemical industry were the pillars of Beijing. However, substantial structural change has taken place since early 2000s. The local extraction of biomass, fossil fuels, and metals decreased significantly, as all material intensive sectors, such as agriculture, steel industry, and chemical industry, declined or moved outside. The resources burden to local natural system has declined over the last decades.

(2) Key challenges at regional level

However, as the population growing quickly, overall consumption of materials kept increasing. The industrial upgrading and restructuring had induced new industries with

① Note: Adapted from the reference [42].

higher value added moving in, such as mobile phones, automobile, etc. Generally, the city increasingly relied on materials imported from outside, which led to a constant growth of local pollution emissions, for example, the air pollution from fossil fuels combustion, as well as the pressing demand on disposal of waste water and municipal solid waste.

(3) Policy priorities for regional circular economy

In general, the resource efficiency in Beijing had been improved continually from 1992 to 2014. The recycling rate of industrial waste improved significantly through growing environmental investment and technological innovation. The stress of linear lock-in had shifted to the consumption end, such as increasing fossil fuel consumption in private transportation, growing waste from consumers. The dependence on resources imported from outside requires the regional close-loop to be constructed beyond the territory of Beijing to include those regions on supply chains. Regional coordination should be the policy priorities in tackling the challenge in the development of regional circular economy.

(4) Designing regional infrastructure for circularity

Significant potential for improvement of circularity exists by improving the low carbon infrastructure in Beijing. The provision of infrastructure for circularity is not only the responsibility of the municipal administration, but it is also an opportunity to create niches for new business models via regional cooperation. The emergence of new technologies and business models can foster the cooperation among stakeholders that cross the border of regions, in which Beijing has the endowment to take the lead.

3.7 Further reading

[1] Cullen J M, Allwood J M. Mapping the global flow of aluminum: from liquid aluminum to end-use goods[J]. Environmental Science & Technology, 2013, 47 (7): 3057-3064.

[2] Miller T R, Duan H, Gregory J, et al. Quantifying domestic used electronics flows using a combination of material flow methodologies: a US case study[J]. Environmental Science & Technology, 2016, 50 (11): 5711-5719.

[3] Krausmann F, Schandl H, Eisenmenger N, et al. Material flow accounting: measuring global material use for sustainable development[J]. Annual Review of Environment and Resources, 2017, 42 (1): 647-675.

[4] Laner D, Zoboli O, Rechberger, H. Statistical entropy analysis to evaluate resource efficiency: phosphorus use in Austria[J]. Ecological Indicators, 2017(83): 232-242.

[5] Jiang D, Chen W Q, Zeng X, et al. Dynamic stocks and flows analysis of Bisphenol A (BPA) in China: 2000—2014[J]. Environmental Science & Technology, 2018, 52 (6): 3706-3715.

[6] Krausmann F, Lauk C, Haas W, et al. From resource extraction to outflows of wastes and emissions: the socioeconomic metabolism of the global economy, 1900—2015[J]. Global Environmental Change, 2018(52): 131-140.

[7] Schandl H, Fischer-Kowalski M, West J, et al. Global material flows and resource

productivity: forty years of evidence[J]. Journal of Industrial Ecology, 2018, 22 (4): 827-838.

[8] Zeng X, Ali S H, Tian J, et al. Mapping anthropogenic mineral generation in China and its implications for a circular economy[J]. Nature Communications, 2020,11(1): 1544.

[9] Di J, Reck B K, Miatto A, et al. United States plastics: large flows, short lifetimes, and negligible recycling [J]. Resources, Conservation and Recycling, 2021 (167): 105440.

[10] Dong D, Tukker A, Steubing B, et al. Assessing China's potential for reducing primary copper demand and associated environmental impacts in the context of energy transition and "Zero waste" policies[J]. Waste Management, 2022(144): 454-467.

3.8 Exercises

1. Please search some literature databases like Web of Science or CNKI to understand how about MFA employs in metal sustainability and its trend.

2. For a certain of substance, there are a number of publications to address its flow analysis. Select any one substance to review all the publications and draw one MFA map. This map can illustrate the evolution of substance flow analysis from time and regions. Please discuss the potential reasons.

3. Please describe the basic characteristic and principle of regional metabolism. Review one city to briefly illustrate this process.

4. Plastic production has been increasing rapidly in recent decades and globally has reach 495 Mt in 2020. Around 4.5% of global greenhouse gas emissions were caused by plastic production and consumption. The massive use of plastic also leads to a large waste, and it was estimated that an accumulative volume of 6.5 billion tons of waste had been generated globally, by the end of 2018. Please review plastic flow analysis in recent years in one country or globally.

5. Main Report: please choose any one topic like element, metal, product, and waste flow to tell one story. And then please write one paper report and prepare one PPT. The paper report: not less than 5 000 words; PPT: 15～20 slides. Please give the presentation in four weeks. The time for you is about 15 minutes. Please submit the final report in six weeks.

Chapter 4 Emerging Methods from Material Flow Analysis

Industrial ecologists have worked to incorporate the MFA approaches into the lifecycle management of product, and combine the monetary flows in evaluation of resource efficiency through cost benefit analysis for decision in real world business. New methods growing from MFA are emerging to deal with the complex relations between market competition and various environmental performances of products. This chapter will present the development of MFA in these fields.

4.1 Life cycle management

Life cycle management (LCM) is important in product development, which is central to the competitiveness of a company in pursuit of sustainability. The life cycle thinking (LCT) is crucial in optimization of the environmental performance while addressing other goals in product development, such as quality, cost, time-to-market, and compliance to regulations. These multi goals are inter-related in the product-oriented material flow management.

4.1.1 Concept from life cycle thinking

LCM is a product management system aiming to minimize environmental and socio-economic burdens associated with an organization's product or product portfolio during its entire life cycle and value chain. LCM makes life cycle thinking and product sustainability operational for businesses to continuously improve the product system.

Life cycle thinking is crucial to sustainable development. It goes beyond the traditional focus on end-of-pipe solution for environmental pollution on production site and manufacturing processes, and incorporates the environmental, social, and economic impact in a systematic way. The environmental management measures and policies that adopt life cycle thinking, such as preventive pollution control, clean production, Extended Producer Responsibility (EPR) and Integrated Product Policies (IPP), require the producers to be responsible for reducing the environmental burdens of their products from cradle to grave. Accordingly, the producers should systematically improve the performance at all stages of the product life cycle.

Figure 4 - 1 shows the lifecycle thinking of a product system that intends to reduce the product's natural resource use and pollutions emissions to the environment, as well as improving its socio-economic performance throughout its life cycle. By considering all the stages of the product life cycle, the designer can ensure that improvements in one stage are not creating bigger problems to another stage of the life cycle. Using life cycle management, the company will see the influence of their choices with regard to sustainability in a more systematic way, balancing the trade-offs among different dimensions regarding the economy, the environment and society. Novel approaches and stakeholder collaboration across the value chain using a full lifecycle analysis approach will be critical for developing the partnerships and circular economy solutions that will lead to scalable implementation.

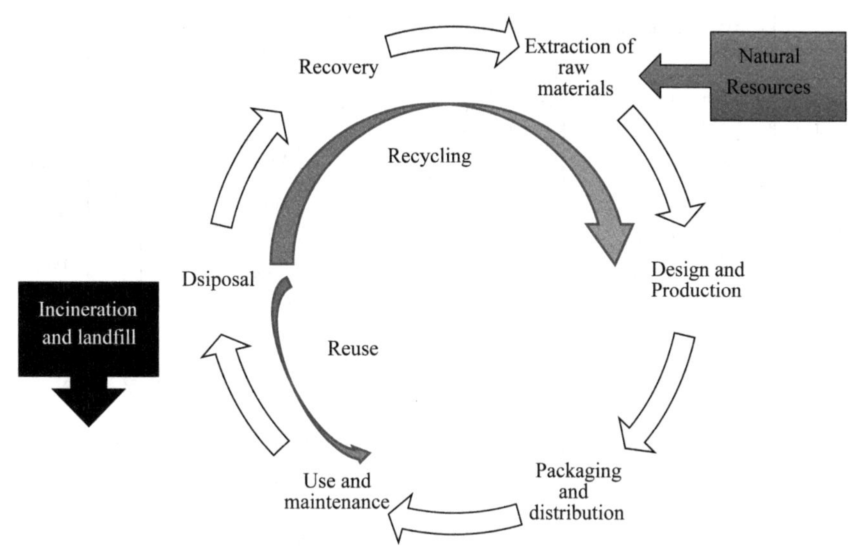

Figure 4 - 1 The lifecycle thinking of a product system starting from extracting raw materials from natural resources in the ground and generating energy(Materials and energy are part of the production, packaging, distribution, use, maintenance, and eventually reuse, recycling, recovery or final disposal) ①

4.1.2 MFA as fundamental tool for life cycle management

LCM provides the integrated framework of organizational sustainability management. The framework connects different concepts, including cleaner production, preventive pollution control, sustainable consumption and production. Various methods, tools and data are used to incorporate environmental, economic and social aspects in the business operation and product development. Among all the tools, the MFA provides the fundamental basic facts about the resources consumption, pollution emission, and waste discharge of the whole product system. The information should be shared among stakeholders, thus to develop shared criteria for product enhancement and value creation with in industry, and ultimately to enhance the company's performance in market competition. Figure 4 - 2 shows the wide range of methods, tools and concepts that can be used in LCM.

1. Incorporate MFA with other analysis tools for life cycle management

LCM generally takes the systematic view for practicable sustainability, helping the

① Adapted from the reference [43].

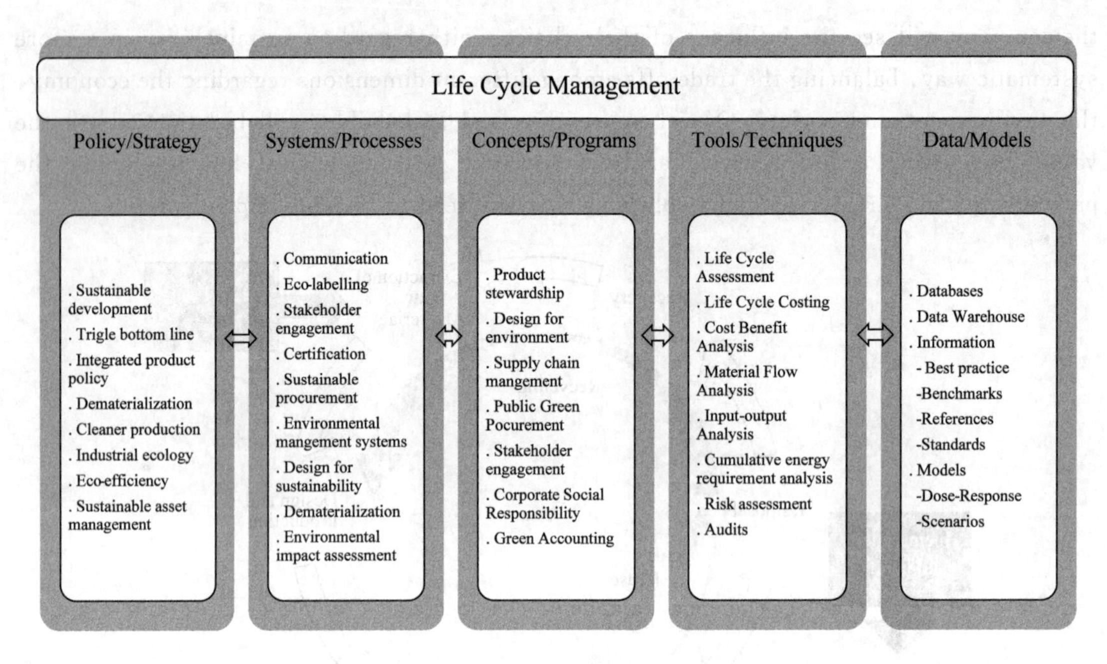

Figure 4 - 2 The policies, strategies, systems, programs, and different types of tools within the life cycle management context①

organization to enhance the environmental performance while achieving other goals in business operation. Various analytical tools are used in LCM to characterize the status of resources use and pollution emission throughout the lifecycle of a product. MFA is the fundamental method for all these analysis tools by quantifying the physical flows of materials through the product lifecycle. The results of MFA provide the scientific base for further evaluation and assessment. For example, different forms of foot printing tools, such as water footprint and carbon footprint, can be used to visualize the scale of environmental burden of the product system calculated from MFA.

The MFA is the essential part of inventory analysis in a series of evaluation tools, such as life cycle assessment (LCA), life cycle costing (LCC), social life cycle assessment (SLCA), organizational LCA (OLCA). With these analysis tools, environmental impacts are integrated with economic and social outcomes in the evaluation (Figure 4 - 3).

In addition, the calculation on MFA at product level provides the basis for the material-based input-output analysis (IOA) on the interrelation across different sectors and regions. For example, the technological change in automobile from fossil fuels to electricity can trigger the change in resource consumption of petrol chemical industry, and economy-wide carbon emission. The precise estimation on these changes relies on detailed MFA on various types of automobile products with different technologies.

① Adapted from the reference [43].

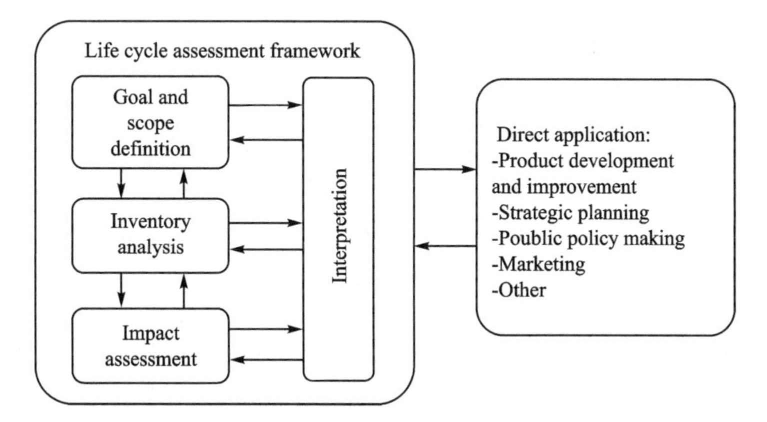

Figure 4 - 3 The standard framework of LCA(MFA is the essential part of the inventory analysis)①

2. Using MFA in operation of LCM

The use of MFA in different types of analysis tools mainly depends on the principal goals of each organization, who uses LCM to support the action for sustainability transition. As to the operations in business, LCM use different procedural tools, including auditing, checklists, and tools of eco-design to support the design for the environment in the stage of new product development. It also helps the actions at cross organization levels, such as brands differentiation through eco-labelling, and improvement of the value chain management.

While addressing policies and strategies of circular economy, LCM incorporated the calculation results from MFA to build indicators of resource efficiency, eco-efficiency, dematerialization for the goals of management, thus coordinate the firm's strategies with the public interests. Life cycle MFA can also be used for the environmental risk assessment (ERA) to assess the risks for any decisions to adjust the material flows in the production processes and operations.

The integrated approach of using MFA with other systems and tools enables organizations to communicate with different stakeholders more efficiently for mutual understanding and cooperative actions in pursuing sustainability, such as communicating with shareholders on the goals and agendas for actions, improving the brand images through life cycle based environmental product declarations to consumers, convincing the public authorities through product information schemes for compliance to green public procurement guidelines, promoting the industry to build market awareness on the products' environmental performance, and collaborating with suppliers with company codes of conduct, audit or supplier evaluation systems to enhance the performance of the entire supply chain. In general, the application of MFA in LCM makes the strategies and actions transparent and visible to the public.

① Adapted from ISO14040.

4.1.3 Implementation of life cycle management

LCM is applicable for industrial and other organizations to implement the preventive and sustainability-driven management in the business. In *Life Cycle Management: A Business Guide to Sustainability* offered by UNEP in 2007, a systematic approach to life cycle management was provided based on the ISO management system standards for environment (ISO 14000) and quality (ISO 9000). The approach addresses the classical management areas, including: ① Policy and goals; ② Action plans and programs; ③ Procedures and instructions; ④ Monitor and revisions; ⑤ Documentation and reporting; ⑥ Communication with stakeholders. The "Plan-Do-Check-Act Cycle" is applied with step-by-step approach to guide the implementation of life cycle management as illustrated in Figure 4 - 4. MFA can be used throughout the cycle.

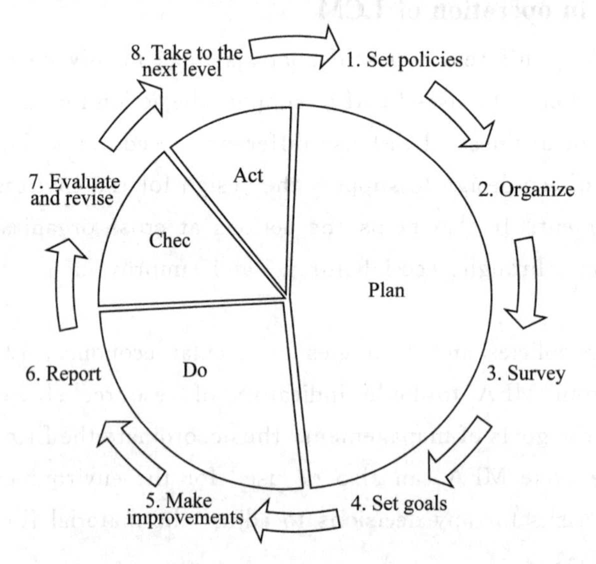

Figure 4 - 4 The phases of step-by-step approach for life cycle management①

1. Plan

Life cycle management has become part of an organization's policies, which need the visionary with a long-range focus for continuous improvements. Under the management frameworks such as ISO 9001, ISO 14001 and/or corporate social responsibility (CSR), MFA can be used to screening the performance of a product from materials selection and acquisition, use of product, distribution to the end-of-life stage. The aim is to identify where and how the LCM process should be started. Based on the environmental profile of the product, the plan for improving the competitive advantages against the competitors is delivered with scientific proof. The plan can be extended to cover suppliers, business associations, authorities, retailers, research institutions, etc. ensures that important aspects are not missed. Thus, concrete goals and the action plan is defined for implementation.

① Adapted from the reference [43].

2. Do

The practical implementation of the plan is vital. MFA provides the tool to guide the organization and process of the plan. The screening of the entire life cycle of a product can identify numerous possibilities for improvements—from the "low-hanging fruit" to systematic change in the long run. The implication of easy improvements will be quick. But the plan to tackle new challenges need to be carefully addressed. For example, when regulations require companies to take back all of their products at the end of life, the company will find that it's difficult to tackle this new challenge by following the traditional business model. It will be appropriate to implement life cycle thinking to engage a redesign of the product to meet the requirements of the legislation, including easier recycling, simple disassembly, new materials selection, etc. In order to track the continuous improvements of the product, MFA should be used in the comparison between different designs in implementation. The information can be used in the corporate sustainability reports to share the progress with the stakeholders.

3. Check

After the implementation of the plan, the evaluation on the experience is needed for policies adjusting. In such an evaluation the MFA can be used to monitor the performance of the processes and products in view of the defined objectives and targets, with the support of indicators from different analysis tools. During this stage, it is essential to ensure the interactions between developing knowledge of a product's environmental and social impacts, market demands, etc. and the implementation of concrete product-oriented improvements. The feedback and criticism from customers and other parties are important for the improvement of products, as well as the product development process.

4. Act

After the periodical review on the progress of the working plan, the possible need for changes will be addressed for policy, objectives and other elements of the system to continuously improve the organization and their products' performance. New round of analysis, including MFA, will be applied to identify areas for further investigations or initiatives, such as preventive and corrective actions as response to potential and actual non-conformities with requirements. If there are significant environmental impacts in the use stage of a product, an investigation of consumers' desires and demands will be necessary. If chemicals or materials, which are on the list of undesirable substances, being used in the product or production process, they should be phased out in the next round action. Throughout the improvement process, the MFA provides the fundamental scientific basis for a product profile, which will support the common sense along the product chain.

4.2 Life cycle cost

Sum of all recurring and one-time (non-recurring) costs over the full life span or a specified period of a good, service, structure, or system. It includes purchase price, installation cost, operating costs, maintenance and upgrade costs, and remaining (residual or salvage) value at the end of ownership or its useful life. With the life cycle thinking dominant in environmental management, the economic aspects have to be incorporated into the evaluation on the improvement of a product, a system, or an activity's environment performance to make the solution economically feasible in the market competition. Life cycle cost (LCC) is an important application of the life cycle management in its economic aspect. This section mainly introduces the development, definition, and framework of LCC.

4.2.1 Concept and framework of life cycle cost in an environmental context

LCC is a process to determine the sum of all expenses associated with a product or project, including acquisition, installation, operation, maintenance, refurbishment, discarding and disposal costs. LCC was developed by the US Department of Defense (DOD) during the mid-1960s and has been used ever since as a tool for large infrastructure projects such as military facilities, buildings, and oil refineries. From the 1980s through the early 1990s, different cost models were developed for LCC estimation, among them, Activity-Based Life Cycle Costing in 2001.

LCC is the sum of the costs throughout the whole life cycle of a product or an engineering project (Figure 4 - 5). Through the LCC assessment, a decision maker can conduct the economic evaluation of a product or an engineering project across its lifetime, thus to choose the best investment plan, based on the least cost. Traditionally, the LCC has been used in cost comparisons over a specific period, considering relevant integral economic factors, including initial costs and future operating costs to optimize product performance and overall cost of ownership. Since 1960s, the LCC methodology has been increasingly used in an environmental context, which is defined as the accounting for all internal and external costs associated with a process or product during its life cycle (Figure 4 - 6).

Many LCC models and methods have been developed over the years as a tool to support the economic decision making. Life cycle cost models can be categorized into three categories which are: ① Conceptual models: This category is for macro-level. It is flexible and based on qualitative variables and hypothesis approach; ② Analytical models: Mathematical models in this category are ranging from simple to very complex models, and they are considered as the most commonly used cost models; ③ Heuristic models: The model can involve simulation models, but cannot guarantee to give an optimum solution. These models are based on ill-structured version of analytical models: an experience-based method and rule-of-thumb strategy, for example, a simulation technique that determines the cost-effectiveness of different levels of reliability and maintainability training for airlines.

The LCC associates the specific categories of environmental impact defined in the

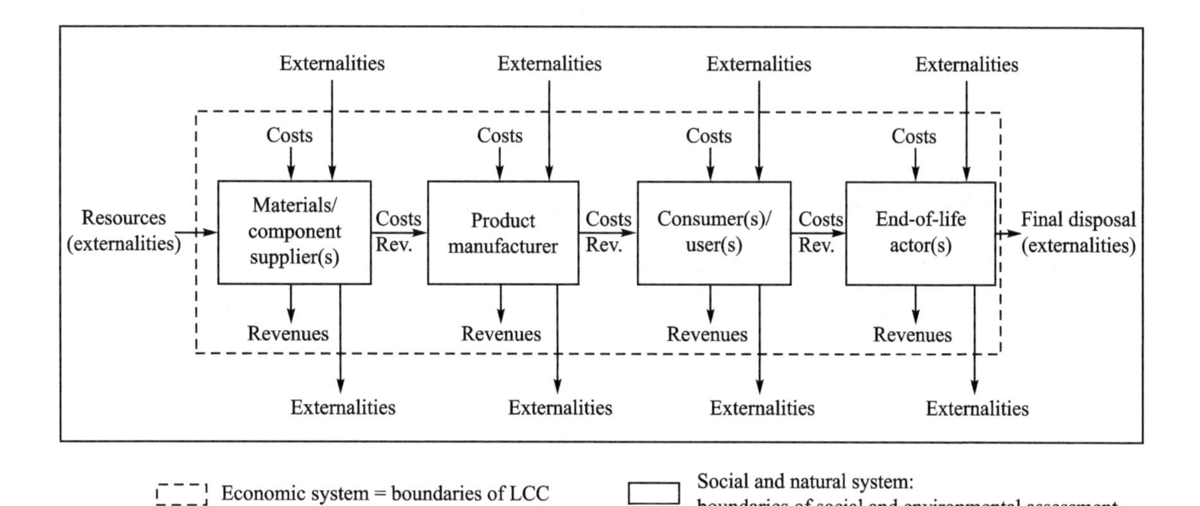

Figure 4 - 5 The conceptual framework of LCC

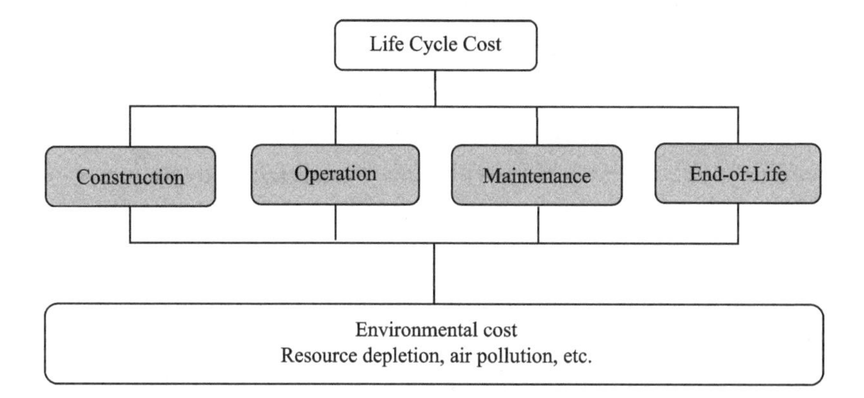

Figure 4 - 6 Components of the life cycle cost in an environmental context[①]

standard life cycle analysis with the accounting of the main costs incurred during the lifetime of the product, work or service, including the internal cost and external cost. The internal costs are those being calculated in the price of a product, such as investment in construction of facilities and equipment, payment to labor forces, cost for maintenance and disposal of EoL waste:

① Construction costs and all associated costs, including delivery, installation, insurance, etc.

② Operating costs, including energy, fuel and water use, spares.

③ Maintenance costs, including the cost of maintenance, repair, and upgrading.

④ EoL costs, such as decommissioning, recycling or disposal.

The external costs are those that cannot be fully calculated in the price of the product, such as resource depletion in the long run, air pollution emission and water contamination,

① Adapted from the reference [45].

etc. The external costs will not be fully covered in the private transactions, and add burden to the society. This is called externality. Special rules are applied for assigning costs to environmental externalities to ensure the cost can be paid by the polluters, and provide the incentives to reduce the pollutions.

4.2.2 The calculation of life cycle cost

The calculation of LCC needs consider all the categories of cost listed in Figure 4 - 5. The calculation should be as comprehensive as possible. The American Society of Testing and Materials (ASTM) published the *Standard Practice for Measuring Life-Cycle Costs of Buildings and Building Systems*, which denotes the equations for LCC calculation:

$$\mathrm{PVLCC}(t,i,C)=\sum_{t=0}^{N}\frac{C_t}{(1+i)^t} \tag{4-1}$$

$$\mathrm{PVCC}=\mathrm{IC}+\mathrm{PV}(\mathrm{OM\&R}+R+F-S) \tag{4-2}$$

$$i=(1+r)(1+I)-1 \tag{4-3}$$

where PVLCC is the present value of the LCC; C_t is the sum of all relevant costs in time t; t is the period of time; N is the duration of the study period; i is the nominal discount rate; PVCC is the main components of LCC; IC is investment costs; PV is the present value (obtained through the nominal discount rate i); OM&R is operating, maintenance, and repair costs' nominal discount rate; R is replacement costs; F is depletion resource costs; S is the residual value; I is the increasing rate; and r is the discount rate.

For the product with long lifetime, such as building, automobiles, and the durable appliances, the discount rate is essential in the calculation for the life cycle cost. However, when the calculation is applied on product with short lifecycle, such as mobile phones, packages, the discount rate can be neglected, thus, the LCC calculation will be simplified as

$$\mathrm{LCC}=\mathrm{IC}+\mathrm{OM\&R}+R+F-S \tag{4-4}$$

The categories of costs can be defined according to the projects to be studied. Using the mathematical tools in LCC can help to identify the economic-sustainable opportunities together with the life cycle assessment on environmental impacts. The results show the direct impact on the economic efficiency of the environmental protection measures, as well as the indirect effect on economic performance.

The LCC calculation results can also help consumers to make good decisions if the producers reveal the financial advantage of an environmentally sound product quantitatively. For example, the environmentally sound products have better performance in LCA, but more expensive. However, if the cost for pollution emission and the high quality for longer usage are considered, the overall cost will be more competitive. In general, assessment of environmental performance with the cost of life cycle is vital in corporate sustainability. Proper public policies informed by LCC will support products both economically and environmentally sustainable in the whole life cycle.

4.2.3 Case study: Life cycle cost analysis for recycling high-tech minerals from waste mobile phones in China

High-tech minerals, including rare earth metals and platinum group metals, are used intensively in manufacturing of mobile phones. As the world's largest mobile phone market, China witnessed dramatic growth in the use of high-tech minerals, which led increasing accumulation of these metals in the waste mobile phone after short lifespan. It is important to investigate the LCC of recycling high tech minerals from waste mobile phones.

In order to recycling the high-tech minerals from the waste mobile phones, the first stage is to collect the waste phones that have already completed their usage stage, and transport to a dismantling facility for recycling. After chemical extraction and treatment procedures, the high-tech minerals in the mobile phones can eventually be recovered. Using MFA and LCC assessment, He P et al. analyzed the recycling of waste mobile phones to obtain high-tech minerals, with comparison on two types of mobile phones, i. e. feature phones and smartphones. The treatment cost of WEEE is composed of the expenditure in collection, transportation, dismantling, recycling, and final disposal. Cost in each stage can be calculated separately through investigation. The total cost can be calculated by sum costs of all the stages. The recycling system boundary of high-tech minerals from waste mobile phones in China and the cost range at different stages is shown in Figure 4 - 7.

Using LCC calculations on various high-tech minerals from waste mobile phones, the final results can be estimated as shown in Table 4 - 1. Through this calculation, we can see that the cost of critical metals used in smartphones is much higher than the feature phones. The cost on cobalt takes the largest share in the cost structure, accounting more than 90% of the life cycle cost on metals used in both feature phones and smartphones. Therefore, guarantee the safety of cobalt supply is very important to the supply chain management in mobile phone production. The technological shift from feature phone to smartphone exaggerates the vulnerability of supply chain as the dependency on cobalt supply increasing. Additionally, more kinds of precious metals are used in smartphones than feature phones. With the huge number of smartphones being produced every year, the demands on these precious metals could create pressure to the supply.

Table 4 - 1 The life cycle cost of high-tech minerals in feature phones and smartphones

High-tech mineral categories	Product categories	
	Feature phone (US $ /unit)	Smartphone (US $ /unit)
Cobalt	6.030	10.110
Palladium	0.014	0.024
Antimony	—	0.135
Beryllium	—	0.005
Neodymium	—	0.008
Praseodymium	—	0.016
Platinum	—	0.006

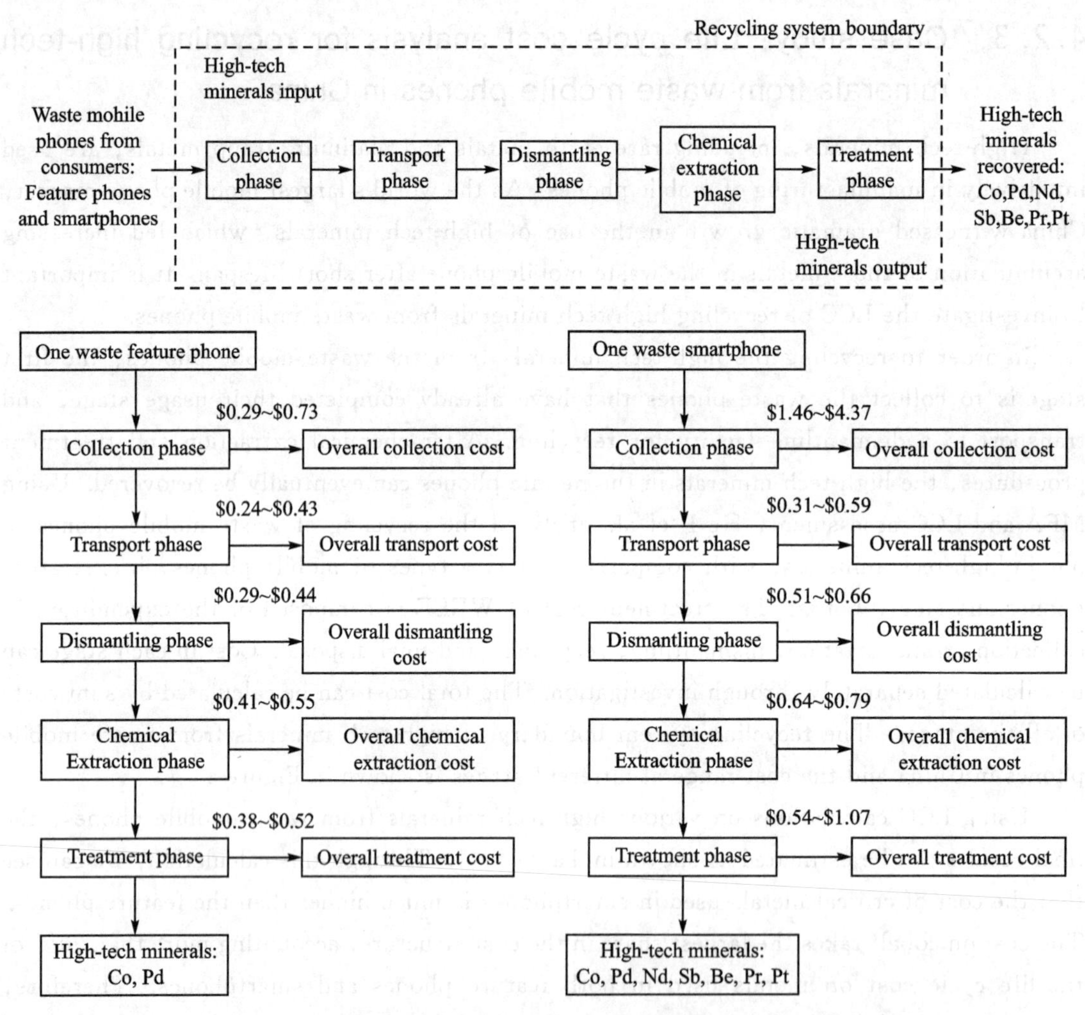

Figure 4 - 7 Recycling system boundary of high-tech materials from waste mobile phones in China and the cost range at different stages①

4.3 Cost-benefit analysis

Value is at the heart of engineering design. It is inevitable to make trade-offs between different design requirements in the development of new product, which represents different values sometimes in conflict with each other. When dealing with trade-offs between environmental performance and other social and economic values, engineers need the optimal design solution with cost-benefit analysis tools.

4.3.1 Concept

Cost-benefit analysis (CBA), sometimes called benefit-cost analysis (BCA), is a systematic approach to estimating the strengths and weaknesses of alternatives that satisfy

① Adapted from the reference [46].

transactions, activities or functional requirements for a business. It is a technique that is used to determine options that provide the best approach for the adoption and practice in terms of benefits in labor, time and cost savings etc. The CBA is also defined as a systematic process for calculating and comparing benefits and costs of a project, decision or government policy.

A CBA is the process for decision making by measuring the benefits of taking an action minus the costs associated with that action. A CBA involves measurable financial metrics such as revenue earned or costs saved as a result of the decision in business operation, including new product development, investment in project, and optimization in processing. As a general method to quantify costs and effectiveness in monetary terms, CBA is useful for comparing very different choices in decision making. It is particularly fitful in the optimization of the expected economic value of a design by maximizing one overarching or super value, such as an economic value like company profits, or the value of the product to users. When conducting CBA, the discount rate, which represent future benefits against current costs (or vice versa), should be considered. The choice of discount rate may have a major impact on the outcome of the analysis.

However, CBA can be controversial if non-economic values are also relevant, which is prevalent in sustainability studies. It is often difficult to indicate what extent values like safety, health, sustainability, and aesthetics contribute to the value of human happiness. Using CBA, means all these intangible values should be expressed in monetary terms for calculation and comparison. In most cases, the subjective value of a person is very different from the actual price on market. In order to reveal the intangible benefits and costs or effects from a decision, approaches like willingness to pay (WTP) have been developed to express values like safety or sustainability in monetary units.

Broadly, CBA has two purposes: to determine if it is a sound investment/decision (justification/feasibility); and to provide a basis for comparing projects. It involves comparing the total expected cost of each option against the total expected benefits, to see whether the benefits outweigh the costs, and by how much. CBA is related to, but distinct from cost-effectiveness analysis. In CBA, benefits and costs are expressed in monetary terms, and are adjusted for the time value of money, so that all flows of benefits and flows of project costs over time (which tend to occur at different points in time) are expressed on a common basis in terms of their "net present value". Closely related, but slightly different, formal techniques include cost-effectiveness analysis, cost-utility analysis, risk-benefit analysis, economic impact analysis, fiscal impact analysis, and social return on investment (SROI) analysis.

Cost-benefit analysis is often used by governments and other organizations, such as private sector businesses, to appraise the desirability of a given policy. It is an analysis of the expected balance of benefits and costs, including an account of foregone alternatives and the status quo. CBA helps predict whether the benefits of a policy outweigh its costs, and by how much relative to other alternatives (i. e. one can rank alternate policies in terms of the cost-benefit ratio). Generally, accurate cost-benefit analysis identifies choices that increase welfare from a utilitarian perspective. Assuming an accurate CBA, changing the status quo by implementing the alternative with the lowest cost-benefit ratio can improve Pareto efficiency. An analyst using CBA should recognize that perfect appraisal of all present and future costs and benefits is difficult, and while CBA can offer a well-educated estimate of the best alternative, perfection in terms of economic efficiency and social welfare is not guaranteed.

The following is a list of steps that comprise a generic cost-benefit analysis: list alternative projects/programs, list stakeholders, select measurements and measure all cost/benefit elements, predict outcome of cost and benefits over relevant time period, convert all costs and benefits into a common currency, apply discount rate, calculate net present value of project options, perform sensitivity analysis, and adopt recommended choice.

4.3.2 Applying cost-benefit analysis with material flow analysis

CBA can be applied with MFA, the central methodology of resource management and industrial ecology in quantifying the materials consumed, recycled, and lost in our society, to improve resource allocation to pursue the maximum economic benefits. The basic idea is to monetize the environmental benefits regarding the material flows throughout the product life cycle and evaluate the overall economic benefits from cradle to grave. There are three kinds of indexes that are often used in CBA to measure the economic benefits, including the economic net present value, the economic internal rate of return, and the benefit-cost ratio. Whatever index is used, there are essential steps in performing the CBA:

① Build the analysis structure: identify the key changes that will be made related to the material flows of a product or processing, and determine all the restraints and requirements clearly.

② Define the stakeholders: who will be impacted by the change, who will bear the costs, and who will obtain the benefits.

③ Categorize the costs and benefits: categorize and classify costs and benefits as direct, indirect, tangible, and intangible in order to determine their effects clearly.

④ Project both costs and benefits: project and evaluate how costs and benefits change over the lifespan of the program/change.

⑤ List the costs as monetary value.

⑥ List the benefits as monetary value.

⑦ Use the index to compare the costs and benefits.

⑧ Perform sensitivity analysis.

⑨ Evaluate the results.

1. CBA with index of economic net present value

The CBA calculated with economic net present value (NPV) takes into consideration the net present value of costs and expected benefits, including the environmental cost associated with the material flows in product lifecycle. This calculation aims to minimize risks and maximize incomes in certain financial period for business operation. After completing the lists of the costs and benefits in monetary value, this index uses net present value to adjust the future cash flows and costs to the present day, for instance, the time value of money concept holds that 1 USD today has a greater value than 1 USD in the future. Therefore, while conducting CBA, the time value of money should be considered in the calculation. Using a discount rate adjusts the future cash flows to the present day. When a change of operation creates current investment in purchasing equipment, and will generate future income flows in following years, the future revenue should be adjusted with discount rate when comparing with the current cost.

The formula to calculate NPV is as following:

$$\mathrm{NPV}=\frac{\text{cash flows}}{(1+i)^{t}}-\text{initial investment} \tag{4-5}$$

where i = discount rate, and t = the number of time periods.

2. CBA with index of economic internal rate of return

The internal rate of return (IRR) is a metric used in financial analysis to estimate the profitability of potential investments. IRR is a discount rate that makes the net present value (NPV) of all cash flows equal to zero in a discounted cash flow analysis. Using IRR as index in the calculations for the cost-benefit analysis of MFA generally relies on the same formula as NPV does. However, IRR is not the actual monetary value. It is the annual return that makes the NPV equal to zero. IRR is fit for analyzing capital budgeting projects to compare potential rates of annual return over time.

The formula and calculation used to determine IRR is as following:

$$0=\mathrm{NPV}=\sum_{t=1}^{T}\frac{C_{t}}{(1+\mathrm{IRR})^{t}}-C_{0} \tag{4-6}$$

where C_t is net cash inflow during the period t; C_0 is total initial investment costs; IRR is the internal rate of return; and t is the number of time periods.

Generally speaking, the higher an internal rate of return, the more desirable an investment is to undertake. IRR is uniform for investments of varying types and, as such, can be used to rank multiple prospective investments or projects on a relatively even basis. Therefore, when comparing options with other similar characteristics, the plan with the highest IRR probably would be considered the best.

3. CBA with index of the benefit-cost ratio

The "Benefit-Cost Ratio" refers to the financial metric that helps in assessing the viability of an upcoming project based on the ratio between its expected costs and benefits. In other words, the ratio determines the relationship between the expected incremental benefit from a project and the corresponding costs that would be incurred to complete the project.

The benefit-cost ratio formula is expressed as present values of all the benefits expected from the project divided by the present values of all the costs to be incurred for the project. Mathematically, it is represented as

$$BCR = PVB/PVC \tag{4-7}$$

where BCR is the benefit-cost ratio; PVB is the present value of the benefits; and PVC is the present value of the costs.

4.3.3 Case study: costs-benefits of virgin and urban mining

In the technological product, copper and aluminum are two of the most widely consumed non-ferrous metals, and their usage and discarding continue to increase. Owing to their unique properties, they play important roles in both basic infrastructure and high-tech sectors such as vehicles, electronics, and new-energy technologies. The containing copper and aluminum are leading in circular economy potential among all the relevant metals. In order to evaluate the economic efficiency for urban mining on copper and aluminum, the costs and benefits analysis is conducted for virgin and urban mining of these two metals regarding recycling of e-waste, end-of-life vehicles (ELVs), and waste wiring and cables (WWC).

As the detailed material and cost flows illustrated in Figure 4 - 8, the cost of recycling old scraps like e-waste, ELVs, and WWC can be obtained. Some CRT TVs are dismantled to obtain valuable resources such as copper, aluminum, iron, and plastic, while others are crushed and separated to obtain copper, iron, and glass. These two procedures yield 0.3 kg and 0.053 kg, respectively, of copper for 1.5 US$ and 3 US$, respectively. Approximately 0.263 kg of aluminum is obtained in each process at 1 US$ (Figure 4 - 8(a)). Likewise, using MFA and LCC tools, the material and cost flows of ELV and WWC recycling are also indicated in Figure 4 - 8(b) and (c), respectively.

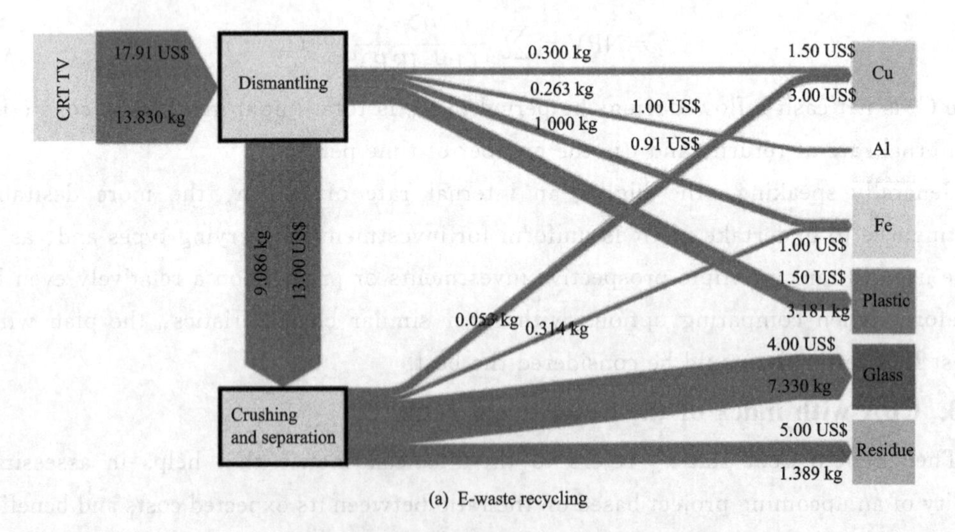

Figure 4 - 8 Material and cost flows of typical urban mining①

① Adapted from the reference [48].

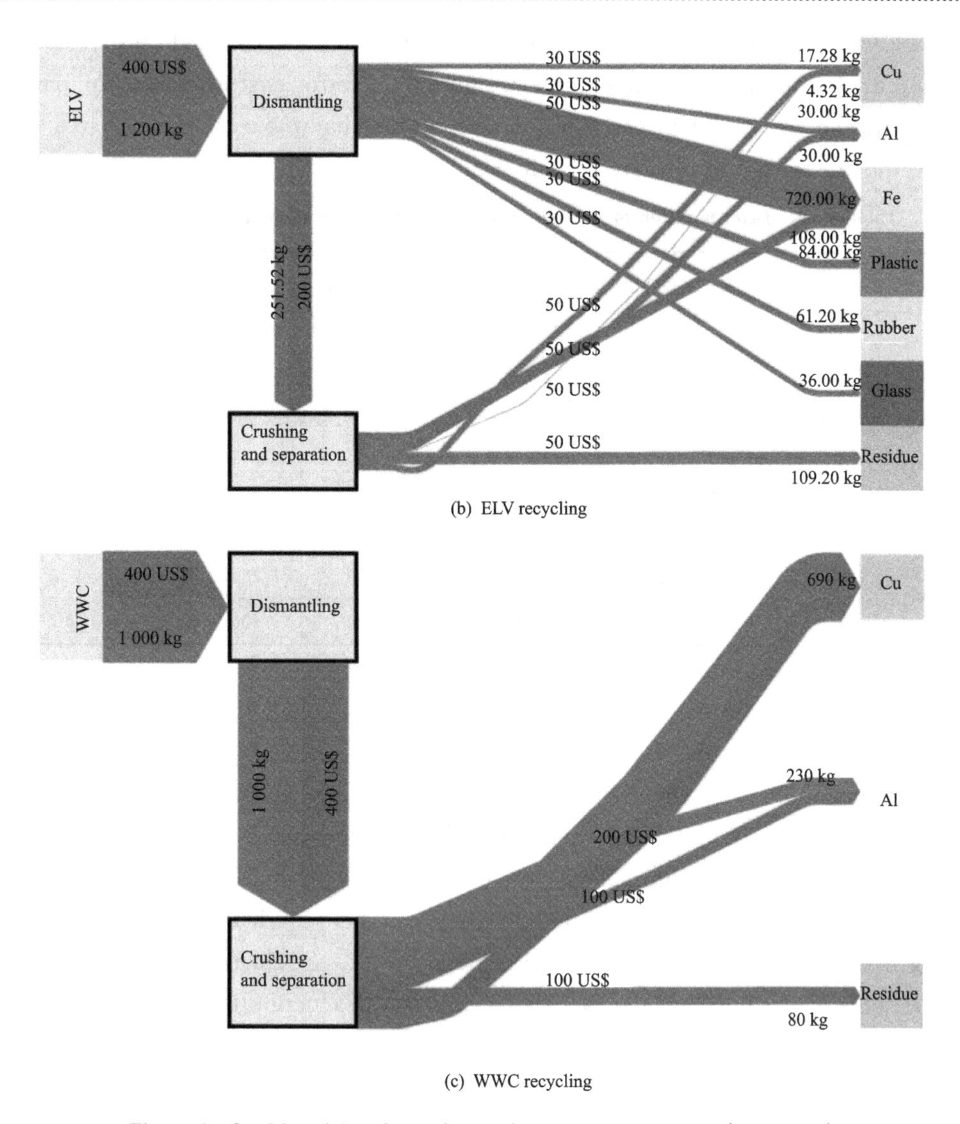

(b) ELV recycling

(c) WWC recycling

Figure 4 - 8 Material and cost flows of typical urban mining(continued)

The urban mining costs for copper and aluminum are listed in Table 4 - 2. In e-waste recycling, the estimated costs for one ton each of copper and aluminum were approximately 12 748 US$ and 3 802 US$, respectively. The recycling costs of ELVs and WWC for the equivalent copper and aluminum amounts are much lower than the cost of e-waste recycling. Thus, the cost of recycling one ton of copper ranges from 290~12 748 US$, while recycling one ton of aluminum will require 435~3 802 US$ (Table 4 - 2).

A comparison of the urban and virgin mining costs and prices demonstrates the economic superiority of urban mining. As complete as possible, data are collected to ensure a comprehensive comparison. The mean and error bar (standard error) are used to demonstrate the differences between values (Figure 4 - 9). For copper, the mean urban mining cost is approximately 3 000 US$ lower than the virgin mining cost (5 500 US$) and price (7 500 US$). Regarding aluminum, the mean urban mining cost is approximately

1 660 US$, which is also lower than the virgin mining cost (2 500 US$) and price (2 200 US$). Therefore, urban mining for copper and aluminum is more cost-effective compared with virgin mining and the metal price. This demonstrates that the economic effects of urban mining are not restricted to e-waste recycling only.

Table 4-2 Determination of the estimated cost for one ton of metal by urban mining

Urban mining	Indicator	Dismantling		Crushing and separation		Estimated cost for one ton of metal* /US$	
		Cu	Al	Cu	Al	Cu	Al
E-waste recycling	Weight/kg	0.3	0.263	0.053	0	12 748	3 802
	Cost/US$	1.5	1	3	0		
ELV recycling	Weight/kg	17.28	30	4.32	50	3 704	750
	Cost/US$	30	30	50	30		
WWC recycling	Weight/kg	0	0	690	230	290	435
	Cost/US$	0	0	200	100		

Note: * The estimated cost for one ton of metal is related to dismantling, crushing and separation. For instance, the estimated cost for copper from e-waste recycling is $1\ 000\times0.3/(0.3+0.053)\times1.5/0.3+1\ 000\times0.053/(0.3+0.053)\times3/0.053=12\ 748$ US$.

Further examination of the costs and benefits of copper and aluminum mining reveals that urban mining allows embodied energy and other resources to be retained compared to virgin mining. As shown in Figure 4-8, the cost of urban mining for aluminum is far below that of urban mining for copper, although they can be recycled with the same process consisting of dismantling, crushing, and separation (i.e. electrostatic or eddy-current isolation). This is attributed to the fact that aluminum production has a much higher energy intensity than copper production. Therefore, its recycling may result in an energy saving of 95%.

CBA indicates cost-efficiency in the perspective of benefit-cost performance. To hunt one-ton copper through virgin mining, hydrometallurgy has over twice the cost efficiency of pyrometallurgy. An interesting finding for urban mining is that recycling from e-waste as CRT TVs and PCBs achieves better benefit-cost performance than those from ELVs and WWC. E-waste, containing heavy metals and toxic organics, has a higher environmental hazard than ELVs and WWC, which needs a high environmental cost. But e-waste recycling can make a rich offset with the subsidy via extended producer responsibility (Zeng et al. 2022). Recycling copper from WWC seems comparatively less cost-effective, but the recycling cost of copper is much higher than that of e-waste. While we further compare urban mining to virgin mining, the benefit-cost ratio of urban mining was on average less than that of virgin mining. The considerable environmental cost is quite indispensable for urban mining with little benefit. Thus, urban mining currently could not achieve cost

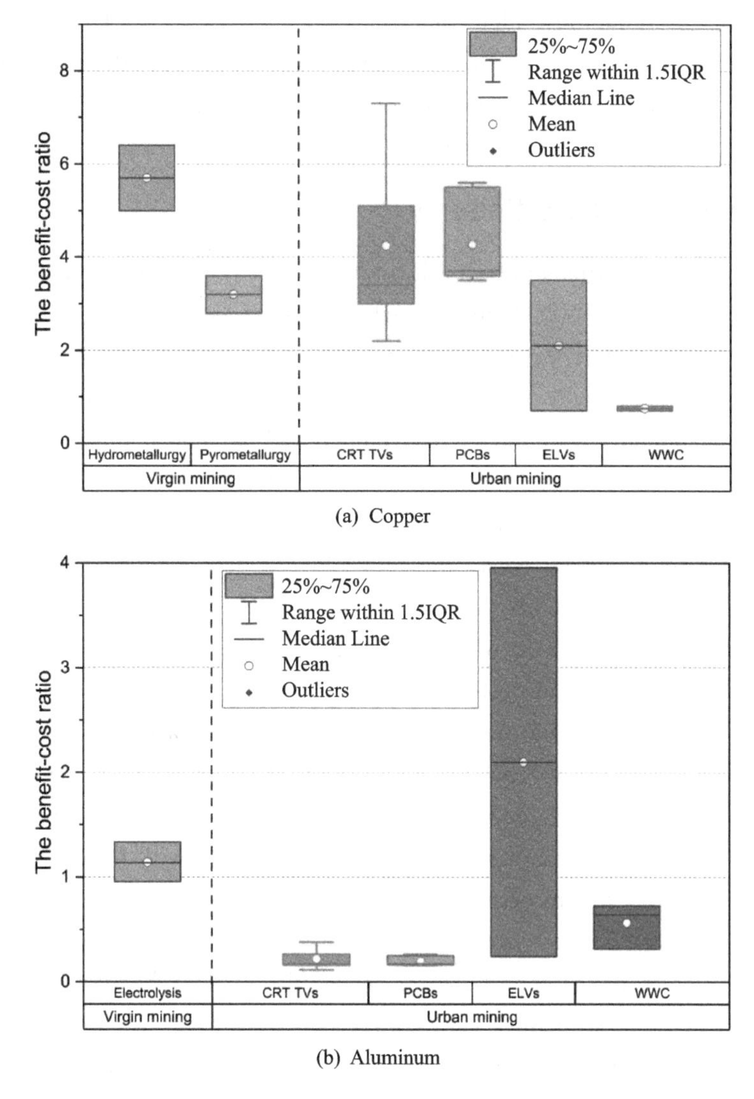

(a) Copper

(b) Aluminum

Figure 4-9 The benefit-cost ratio for the mining process(each range of the error bars was defined based on mathematical statistics for all the estimated data)①

efficiency over virgin mining. To promote urban mining to replace the virgin mining, policies are needed to decrease the cost or raise the benefit.

4.4 Ecological efficiency

In different tools for the assessment of the sustainability impacts and benefits occurring across a product's life cycle, the ecological efficiency analysis provides an integrated framework to facilitate strategic decision making along the entire value chain. It enables companies to focus on sustainable product development with competitive advantage on the

① Adapted from the reference [48].

marketplace. The methodology identifies the factors that can effectively improve the sustainability profile of a product at the very early stage of product development. With clear understanding of trade-offs, the concept of ecological efficiency aims to prevent the inadvertent shifting of environmental impacts from one area to another.

4.4.1 Concept of eco-efficiency

The concept of eco-efficiency was introduced by the World Business Council for Sustainable Development (WBCSD) in the early 1990s. It is a management philosophy that encourages business to search for environmental improvements that yield parallel economic benefits. It focuses on business opportunities and allows companies to become more environmentally responsible and more profitable, and is a key business contribution to sustainable societies. As defined by WBCSD, "eco-efficiency is achieved by the delivery of competitively priced goods and services that satisfy human needs and bring quality of life, while progressively reducing ecological impacts and resource intensity throughout the life-cycle to a level at least in line with the earth's estimated carrying capacity." In short, it is concerned with creating more value with less impact.

The concept of ecological efficiency has moved from preventing pollution in manufacturing industries to becoming a driver for innovation and competitiveness. Companies implement ecological efficiency to optimize their processes, turn their wastes into resources for other industries, and drive innovation that leads to products with new functionalities. It is an integral part of a strategy, which will have a strong focus on technological and social innovation, accountability and transparency, as well as on cooperation with other parts of society with a view to achieving the set objectives (Figure 4 - 10). Thus, the concept of eco-efficiency makes good business sense in improving business and environmental performance, and helping companies gain competitive advantage for long-term profitability and sustainability. Additionally, eco-efficiency can scale up to the entire supply and consumption production value-chain, and achieve macroeconomic efficiencies for countries by promoting the qualitative growth rather than quantitative one that consuming more resources and generating more waste for the same value and function production.

Eco-efficiency has been implemented along the entire value chain of a product or service, beyond the boundaries of the manufacturing plants. For companies with extensive supply network, the most critical environment burdens could occur outside the companies - either upstream in the raw material production and components suppliers, or downstream in the product use or disposal phases. In order to achieve the systematic optimization of the entire product system, eco-efficiency can be achieved with seven key approaches (REDUCES):

① Reduce material intensity.

② Energy intensity minimized.

③ Dispersion of toxic substances is reduced.

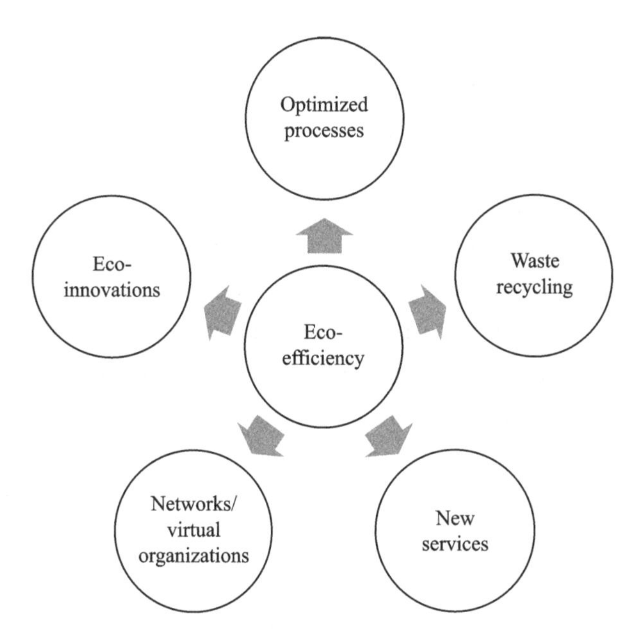

Figure 4 - 10 Five aspects of eco-efficiency as indispensable strategic elements in business①

④ Undertake recycling.

⑤ Capitalize on use of renewables.

⑥ Extend product durability.

⑦ Service intensity is increased.

4.4.2 Measuring ecological efficiency with material flow analysis

Implementing eco-efficiency along the entire value chain of a product or service entails various techniques and tools, which range from simple industrial practices related to improving resource and energy efficiency to highly innovative product and process redesign initiatives, in which ecological or environmental considerations are used as a catalyst for change. Materials flow analysis can be used to identify benefits of eco-efficiency for industrial production system, which involves energy and materials usage in production, combined with other tools, including LCA and materials flow cost accounting.

In order to measure the eco-efficiency performance to track performance and progress and identify opportunities for improvement, a basic equation can be used to indicate the essence of this concept—providing more value per unit of environmental influence or unit of resource consumed:

$$\text{Eco-efficiency} = \frac{\text{Product or Service value}}{\text{Environmental Influences}} \tag{4-7}$$

This equation brings together the two eco-dimensions of economy and ecology to relate product or service value to environmental influence. The material flows of resources and energy consumed, as well as waste emission are associated with the environmental

① Adapted from the reference [50].

influences. While, the product or value represents the economic outputs. The calculation can be conducted for different entities, such as production lines, manufacturing sites, or entire corporations, as well as for single product, market segments, or entire economies. The results can be compared among different companies, products, and economies to create credibility among investors and stakeholders.

The WBCSD developed a common framework for eco-efficiency indicators (Figure 4 - 11), with terminology consistent with the ISO 14000 series and the Global Reporting Initiative (GRI). They define three levels of organization for eco-efficiency information—categories, aspects and indicators: categories are broad areas of environmental influence or business value. Each has a number of aspects, which are general types of information related to a specific category. Aspects describe what is to be measured. Indicators are the specific measures of an individual aspect that can be used to track and demonstrate performance. A given aspect may have several indicators.

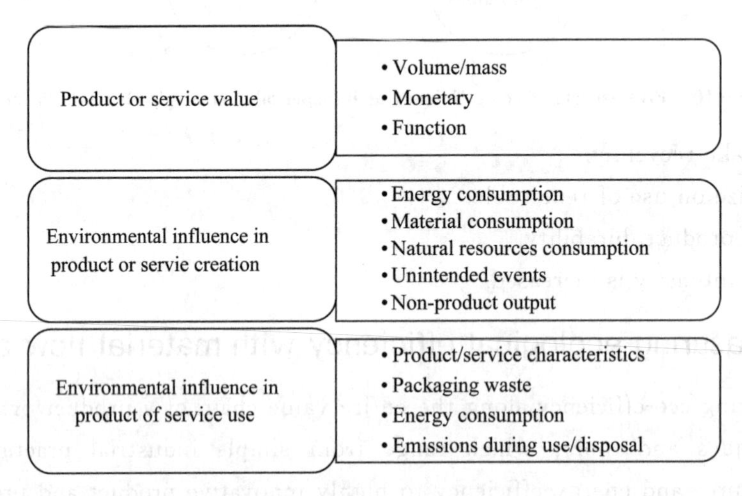

Figure 4 - 11 Three categories of eco-efficiency measurements and their main related aspects①

The indicators fall into two groups. The generally applicable indicators for product or service value are quantity of goods or services produced or provided to customers, and net sales. The indicators related to the environmental influence in product or service creation are energy consumption, materials consumption, water consumption, greenhouse gas emissions, and ozone-depleting substance emissions etc. The material flow analysis provides scientifically support information on the environmental influences in these indicators, which provide accurate and comparable information in the measurement of eco-efficiency for business decision making.

4.4.3 Case study: eco-efficiency analysis of BASF

The concept of eco-efficiency provides a powerful tool to enterprises to integrate the idea of sustainability into their business strategy. Many companies developed their own methods

① Adapted from the reference [50].

and analysis tools in their business operation. For example, BASF, the world leading chemical enterprise, has developed a user-friendly model to support their customers' sustainable development along the value chain by reducing energy and resources since 1996. In order to be harmonious with other business operations in practice, the model follows ISO 14040:2006 and 14044:2006 for environmental life cycle assessments. And the assessment of life cycle costs and aggregation to an overall eco-efficiency is based on ISO 14045:2012.

The eco-efficiency analysis of BASF compares the life cycles of products or manufacturing processes in a holistic approach from the assessment including raw materials sourcing, product manufacture and use, to disposal or recycling. The calculation includes not only the environmental impact of products used by BASF, but also materials manufactured by other suppliers. The analysis considers the consumption behavior of end-users and compares the environmental performance of different recycling and disposal options (Figure 4 - 12). For example, when Comexi, a flexible packaging solution provider, wanted to find out which laminating adhesive technologies are the most eco-efficient, they use the BASF eco-efficiency analysis model compares the alternative technologies, including solvent-based polyurethane (SB), solventless polyurethane (SL), water-based polyurethane (WB-PU) and water-based acrylic (WB-A) laminating adhesives. The analysis can answer the question like how much energy is needed to fulfill the customer benefit, what emissions and waste result, given required packaging parameters and costs constrains. Thus, help the company to find the most eco-efficient solution for their customers.

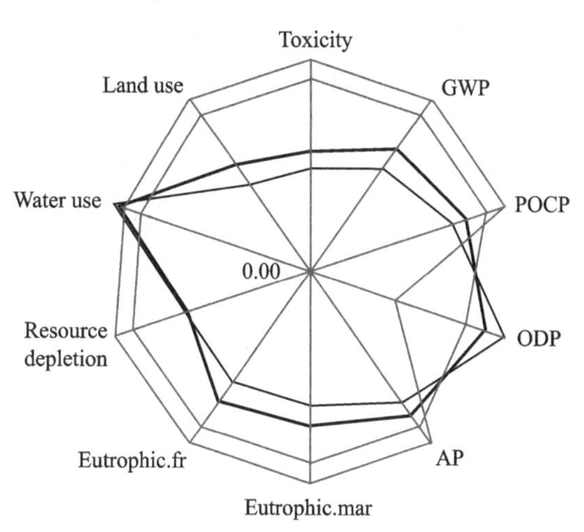

Figure 4 - 12 The environmental fingerprints for different solutions in comparison[1]

The environmental impact is assessed in BASF's eco-efficiency analysis model based on a range of categories[2]:

① Raw materials consumption (Resource depletion).

② Water consumption (Water use).

③ Land use.

④ Human toxicity potential (Toxicity).

⑤ Eutrophication (Eutrophic. fr, Eutrophic. mar).

⑥ Acidification (AP).

---

① Adapted from BASF's official website (https://www.basf.com/).

② https://www.basf.com/global/en/who-we-are/sustainability/we-drive-sustainable-solutions/quantifying-sustainability/eco-efficiency-analysis.html.

⑦ Ozone depletion (ODP).

⑧ Photochemical ozone creation (POCP).

⑨ Climate change (GWP).

Each category of environmental impact will be calculated based on the material flow analysis on the materials in concerns used in production. The cost and revenue data associated with the material flows are also compiled. Thus, the economic analysis and the overall environmental impact can be complied into an integrated framework to make eco-efficiency comparisons between different solutions.

BASF method provides an eco-efficiency portfolio metrics to visualize the results calculated from the economic and ecological data (Figure 4 - 13). The costs are shown on the horizontal axis and the environmental impact is shown on the vertical axis. The graph reveals the eco-efficiency of a product or process compared to other products or processes. And it can be used in scenario analysis for making strategic decisions and detecting opportunities for ecological and economic improvements.

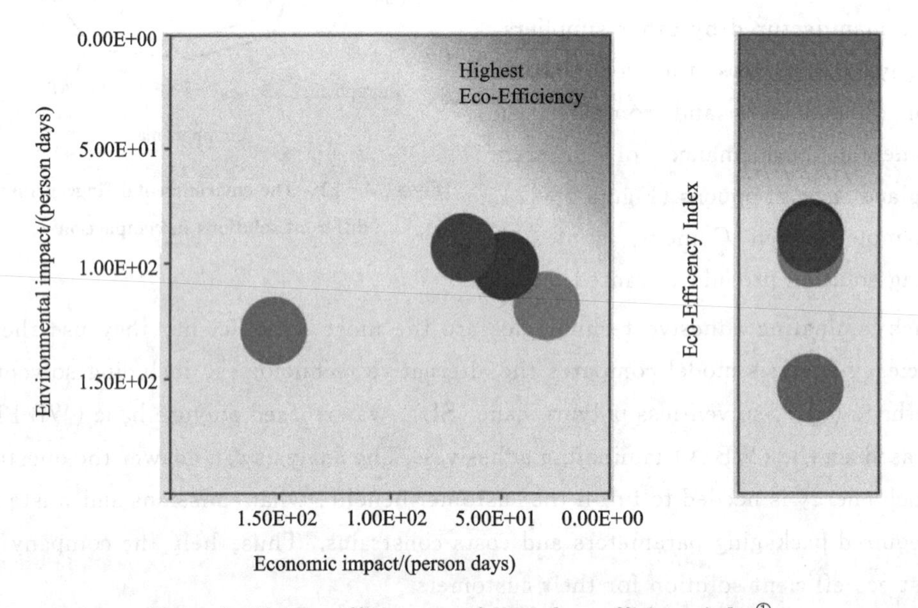

Figure 4 - 13 Eco-efficiency portfolio and eco-efficiency index[①]

The BASF's eco-efficiency analysis model has been adopted by other organizations in policy making. For example, the Association of Plastic Manufacturers in Europe (APME) conducted an integral analysis on the relative effectiveness in economic and environmental impact of various plastic waste recovery strategies. The calculation used the BASF's eco-efficiency model to present an overview of the environmental aspects and economic impacts of actual reference scenarios and different theoretical scenarios of "state of the art" processing routes of packaging plastics in Europe, including collection, pre-processing, mechanical recycling, feedstock recycling, energy recovery and residues incineration. The aim was to

① Adapted from BASF's official website (https://www.basf.com/).

illustrate the eco-efficiency that could be achieved through the improvement in the plastic packaging waste processing in the discussion with stakeholders and policy makers, and supporting the impending revision of the targets in European Packaging and Packaging Waste Directive.

4.5 Statistical entropy analysis

4.5.1 Concept of statistical entropy

Entropy is a scientific concept as well as a measurable physical property that is most associated with a state of disorder, randomness, or uncertainty. It is an extensive property of a thermodynamic system, which means its value changes depending on the amount of matter that is present. Given equal probability of outcomes, entropy equals Boltzmann's constant (k_B) multiplied by the natural logarithm of the number of possible states (W): $S = k_B \ln W$, where Boltzmann's constant is $1.380\ 65 \times 10^{-23}$ J/K.

Qualitatively, entropy is simply a measure how much the energy of atoms and molecules become more spread out in a process and can be defined in terms of statistical probabilities of a system or in terms of the other thermodynamic quantities. The second law of thermodynamics states the total entropy of a closed system cannot decrease. However, within a system, entropy of one system can decrease by raising entropy of another system.

The term and the concept are used in diverse fields, from classical thermodynamics, where it was first recognized, to the microscopic description of nature in statistical physics, and to the principles of information theory. It has found far-ranging applications in chemistry and physics, in biological systems and their relation to life, in cosmology, economics, sociology, weather science, climate change, and information systems including the transmission of information in telecommunication.

The concentration of all materials in product has been substantially evaluated by Shannon's Statistical Entropy function of Information Theory, which can measure the loss or gain of information about a system or the variance of a probability distribution. This theory has been successfully adopted in material flow assessment through quantifying the power of a system to concentrate or dilute substances.

4.5.2 Measuring the recyclability of product waste

As a global society we are still far from a closed-loop materials system, more efficient recycling is crucial to enable the recovery of critical materials and solve the global waste problem. Basically, efficient recycling for product waste is dependent on sophiscated technology and processes. A theoretical guide to recycling process is of necessity to utilize the experience and to cope with emerging electronics. Additionally, eco-design (or design for environment) has been much emphasized by producers, consumers, and recyclers. But extensive analysis of the eco-design of products by many producers has determined that such design is currently focused only at the qualitative level. Quantification of eco-design is therefore also needed for producers to ensure that their new product is indeed superior to the old ones.

Product recycling is much concerned by producers or manufacturers under the EPR policy—in most cases, via paying the recycling cost, which is strongly related to the economic value of recycled products and the overall processing cost. The former is easily obtained from the composition of product waste. The economic value of materials contained in product waste depends not only on the concentration of recyclable materials contained in the products, but also upon the chemical property (e. g. valence) of the materials. Thus, the overall expenses of product waste treatment are strongly related to the full recycling process, including physical treatment (e. g. manual dismantling, mechanical treatment, or thermal treatment) and chemical recovery (e. g. hydrometallurgy and pyrometallurgy).

The term"recycling potential" has been defined from an economic perspective, satisfying the requirement that potential product waste recycling revenues exceed the costs of collection, transportation, and processing. Owing to economic fluctuation and rapid technological innovation, however, recycling potential does not take into account the entire processing cost and thus cannot always reflect the nature of the recycling technology. Here we define recyclability as the theoretical probability of an item's actually being recycled, taking into account the recycling difficulty for metals, plastics and glass in physical treatment and chemical recovery. Therefore, the net recycling cost of product waste depends upon its recyclability, not its recycling potential, which is only relevant to the concentration of recoverable materials within it. A comprehensive theoretical guide to recycling processes, quantitative eco-design, and recycling responsibility is quite essential to determine the authentic recyclability of product waste. Here we create innovative mathematical models, based on the physical and chemical characteristics of materials contained in product waste, to measure the recyclability of various types of product.

1. Method

Aside from external factors such as available technology and economic feasibility, the recyclability of product is in essence dependent upon the types and quantities of its elements and their grades. For instance, metals' recyclability in urban mining is also attributed to metals form: metals used in pure form are judged to be relatively easy to recycle. The types and quantities refer to the concentrations of various valuable materials (e. g. metals and plastics) and other materials, for instance, product waste recycling should consider not only the metals (as elemental or combined substances), plastics and glass, but also some fewer valuable materials such as resin. The grade is to measure the quality of various materials, which varies depending on the levels of natural minerals in a given substance.

The statistical entropy H of a finite probability distribution is expressed by the following equations:

$$H = -\sum_{i=1}^{n} (P_i \cdot \log_2 P_i) \tag{4-8}$$

$$\sum_{i=1}^{n} P_i = 1 \tag{4-9}$$

where P_i is the probability that event i happens, replacing the concentration of i material; n is the total number of materials in a given product waste; and H is the entropy (here its unit is a bit in Information Theory).

Here we define that the initial phase of product waste recycling is the EoL product, and the targeted phase is the final recovered materials. Actually, the final recovered materials are depended upon the utilization purpose, and the recycling & recovery process. But in most cases, the final recovered materials from product waste are pure metals or plastics using physical treatment and hydrometallurgical process. Thus, the targeted phase of product waste recycling can be defined as pure metals or plastics.

The grade of the materials within product waste is the most difficult to determine, but it can be revealed by quantifying the divergence between the initial phase and the targeted phase as recovered product. Basically, two situations can be identified to address this issue. First, in the case of some mixed-material goods, the grade of one material is regarded as the concentration of the goods. For instance, metallic powder containing copper and nonmetals could be commonly obtained from the mechanical and electrostatic separation of product waste; thus, the grade of copper (zero-valence state) in the goods can be determined with the content of copper. The grade of alloy, in particular, will be defined as the concentration of a certain metal based on the assumption that the target recovery phase is pure metal (zero-valence state). Secondly, with respect to some metals in a chemically combined substance, the grade is subject to the combined price. If the targeted phase is pure copper, for instance, the grade of CuO (two-valence state) is lower than the one of Cu_2O (one-valence state). Hence, the grade of goods can be defined as the sum of the grades of all its materials, as shown in Eq. (4 - 10):

$$D = \sum_{i=1}^{m} D_i = \begin{cases} \sum_{i=1}^{m} P_i & \text{(physically mixed goods)} \\ \sum_{i=1}^{m} [1-(j_i-1)/N]_i & \text{(chemically combined goods)} \end{cases} \tag{4-10}$$

where D_i is the grade of material i; m is the number of materials in a given product waste ($m \leqslant n$); j is the ranking of valence from low to high; N is the total number of all the valences of a given element; and D is the total grade of the product waste (here its unit is dimensionless).

In order to demonstrate how this method can be used to classify the grade, some cases related to the two types of goods are presented here, along with their grades, based on Eq. (4 - 10). The grades of typical materials or substances were shown in Table 4 - 3. Because of the diversity and variability of recycling technology, here a basic assumption is that no significant difference based on physical or chemical processes determines a good's viability for product recycling (physical and chemical processes are normally utilized to deal with physically mixed goods and chemically combined ones, respectively). Most goods are composed of both physically mixed goods and chemically combined ones. Therefore, their total grades can be determined by the composite compilation of all their constituent goods.

Table 4 - 3 Classification of materials/substances and their grade determinations

Materials/Substances	Category	Example	Target phase	D
Physically mixed materials	Simple mixed goods	Metallic powder[a]	Cu	0.90
	Alloy	$Sn_{63}Pb_{37}$	Sn and Pb	1
Chemically combined substances	—	CuO[b]	Cu	1/3
		Cu_2O[b]	Cu	2/3

Note: [a] The concentration of copper in metallic powder recycled from waste PCBs is approximately 90%;

[b] j is 1, 2, and 3 for Cu, Cu_2O, and CuO, respectively, and n and N are 1 and 3, respectively.

Clearly, the recyclability will decline with the increase of entropy H. A high grade of D will enhance the recyclability, while complexity will likely lessen the recyclability. Consequently, the recyclability can be defined by Eq. (4 - 11).

$$R = \frac{100D}{n \cdot H} \tag{4-11}$$

where R is the recyclability (/bit), and 100 is coefficient of amplification.

In physics, entropy is a measure of disorder, and high entropy means significant disorder. High recyclability R with a unit of /bit means low disorder, or a homogeneity of the composition, which promises eminent recyclability. As a result, since H is related to mass percentage, the value of R reveals the average recycling possibility of product waste per unit mass, and implies the recycling cost for product. In order to determine the recyclability with some certainty, the boundaries (i. e. minimum and maximum) of the recyclability should be well defined using average values for the earth's crust and for the pure compound, respectively. More substances considered here, more accurate for the recyclability determination. The theoretical recyclability of the earth's crust is less than 18 /bit. According to Eq. (4-8), the entropy H of a goods containing only one material is 0; thus, its recyclability is infinite (∞). All the results indicate that the recyclability of any material is above 18 /bit, and that the recyclability of natural minerals ranges from 29 to 55 /bit with entropy H of 0.97~1.9 bit.

2. Recyclability of various types of product waste

The procedure for determining the recyclability of product waste item can be illustrated using three typical air conditioners as examples. Based on available information about the composition of air conditioners, the statistical entropy H values were determined: 1.5, 1.8, and 2.0 bit. If the targeted phase is pure metal, the grade can be defined as the concentration of dismantled materials (e. g. Al, Cu, and Fe) (approximately 90% of dismantled Al materials are pure). Thus, the total grade for an air conditioner is 3.8. According to Eq. (4 - 11), the recyclability of the three compositions is 50, 42, and 37/bit, indicating that the recyclability of an air conditioner is about (43±5.1)/bit (43 is the mean and 5.1 is the standard deviation).

Similarly, the compositions of many types of products are collected as soon as possible; their grades are also demonstrated that some hazardous materials and precious metals have been considered, but not presented separately owing to their low concentration. Table 4 - 4 summarizes entropy H, grade D, and recyclability R for various types of electronics. The entropy H of each electronic product is expressed within a certain range; variations can be attributed to differences among manufacturers, production dates, and product sizes. Since, however, a given type of product is always composed of similar structures and components, regardless of such variations, that particular product will show a relatively constant grade of D. The overall grade range of electronics is around 3.5~6.8 (Table 4 - 4). A higher value indicates higher purity of each material and a more complex composition. For instance, a printer can be dismantled into seven materials, and the majority of these remain at excellent grade after extraction.

Table 4 - 4 Determination of H, D, and R for various types of electronics

Electronic product	H/bit	D	R/(bit^{-1})
Air conditioner	1.8~1.9	4.1	43±5.1
TV	1.6~2.3	4.2	37±4.9
Refrigerator	1.4~1.7	4.7	54±3.7
Washing machine	1.4~2.0	4.7	48±6.1
CRT DC *	1.4~2.4	3.8	52±3.4
Mobile phone	2.0~3.1	3.5	28±0.99
Duplicator	1.3~1.8	5.1	49±8.5
Pinter	1.7~2.3	6.8	39±5.6
Scanner	2.3~2.4	4.5	32±0.07

Note: * CRT PC is a desktop computer (PC) with a CRT monitor.

Secondary product waste (i.e. WEEE parts) should not be ignored, as they are usually derived from dismantling the larger products. The entropy H, grade D, and recyclability R of WEEE parts are displayed in Table 4 - 4. The entropy H and grade D tend to be within the ranges of 1.5~3.0 bit and 2.4~6.5, respectively. The maximum (6.4) is from printed circuit boards (PCBs) and the minimum (2.5) is from printed wiring boards (PWBs) dismantled from PCBs, because more than sixty elements were found in PCBs, and the dismantling of PCBs is usually required for the comprehensive recycling process.

In order to clearly demonstrate the recyclability, a recyclability map for product waste (including WEEE, WEEE parts, and ELV) is outlined in Figure 4 - 13. Those recyclability ranges from 20 to 60/bit, clearly higher than that of the earth's crust. The H value of PCBs is obviously higher than that of other e-waste, revealing that the physical content of PCBs is more complex. In contrast, washing machines and duplicators present a low H value. More importantly, the complex content of current product results in higher H values than those of natural minerals.

3. Comparison and validation of the related results

The entropy of product recycling was ever determined with some previous studies. The range of entropy also reveals the continual evolution of electronic product. From the R-dimension perspective, the recyclability of product waste can be divided into three levels (Figure 4－14): to the first level ($R>50$) belong desktop computers, washing machines, and some duplicators, which are easy to recycle; the moderate level ($30<R<50$) comprises most types of e-waste, including air conditioners, refrigerators, spent LiBs, toner cartridges, LCD monitors, LEDs, printers, and scanners; and the last level ($R<30$) comprises PCBs, PWBs, and mobile phones, which are the most difficult of all to recycle. These results are to some extent supported by the fact that washing machines can be recycled through simple dismantling and sorting, while PCBs must be treated through a complex process encompassing dismantling, crushing, separation, and hydrometallurgy or pyrometallurgy.

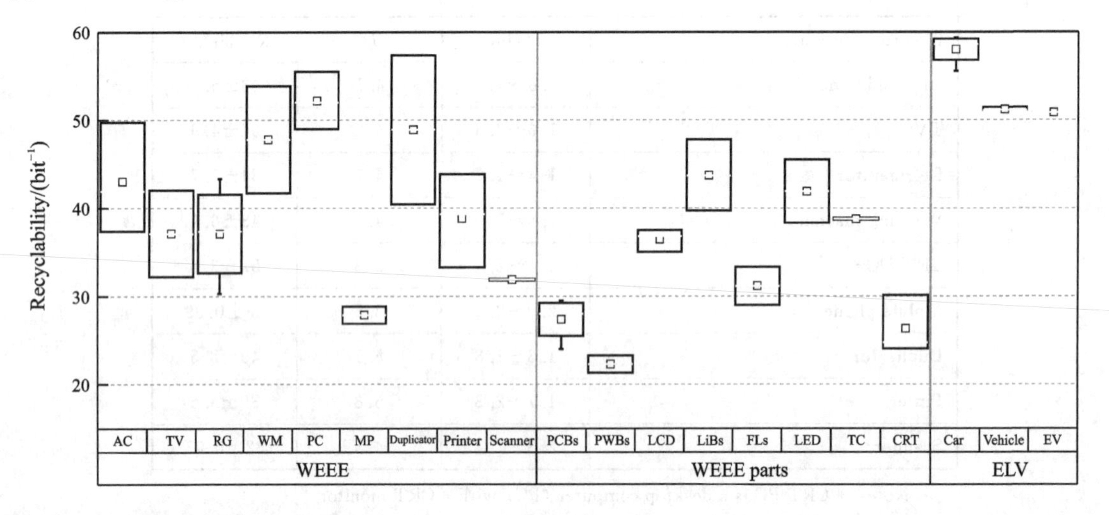

AC—air conditioner; RG—refrigerator; WM—washing machine;
DC—desktop computer; MP—mobile phone; TC—Toner cartridge; CRT—CRT glass.

Figure 4－14 Systematic map for recyclability of product waste. Each bar shows the range of recyclability offered by a material, of which are labeled

4. Sensitivity analysis for the obtained results

The above results were derived from the assumption that pure metals or plastics are the targeted phase of product waste recycling. If the targeted phase is changed as alloy or chemical compound, similar mathematical models can be still employed to measure the recyclability despite some altered parameters.

To be scientific, sensitivity analysis is quite necessary to evaluate the robust of the achieved results and mathematical models. The recyclability of the earth's crust is still chosen for sensitivity analysis owing to more materials or elements in crust. The result of sensitivity analysis is illustrated in Figure 4－15. Although the recyclability declines with the growth of the number of elements, yet the ratio can remain the range of 90%～100%. It can reveal that a relatively high robustness of recyclability occurs when the total composition of calculated materials except for others is over 95%.

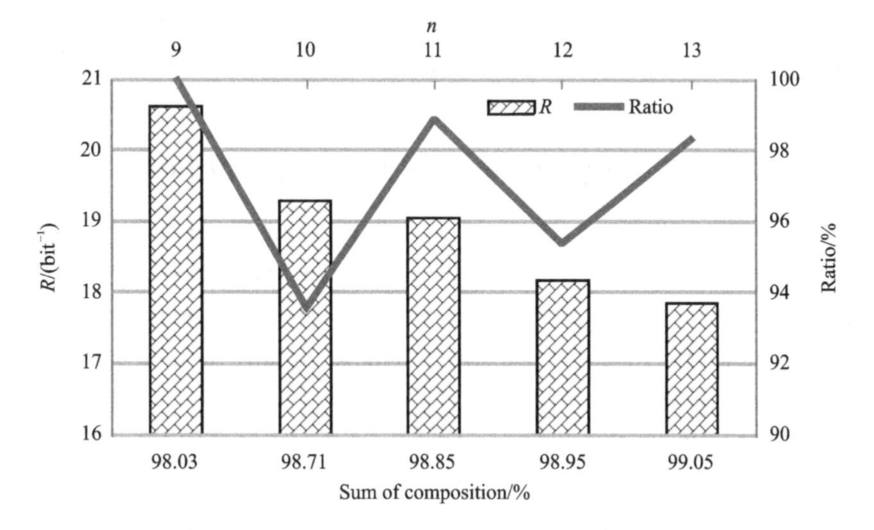

Note: sum of composition is the total composition of elements except others, and the ratio is determined by $(R_{n+1}/R_n)\times 100\%$.

Figure 4 - 15　Sensitivity analysis for recyclability in case of the number of materials

5. Implication for choosing a recycling process

The obtained three recyclability levels of product waste imply diverse options for recycling technology. In the "easy" recycling category, simple dismantling is the major approach; in the "moderate" category, a combined process of dismantling and simple mechanical recycling is required; and in the "difficult" category, vigorous and intense physical treatment and chemical recovery are indispensable. Given this knowledge, more reasonable recycling process for emerging types of electronics can be schematized. The achieved results in Figure 4 - 14 can contribute to choosing a recycling process, even for emerging product waste.

6. Implication for quantifying eco-design

EPR policies can provide producers incentives for eco-design (more sustainable and easily recyclable electronics) as they may be faced with the financial or physical burden of recycling their product after use. Each bar shown in Figure 4 - 13 shows the range of recyclability for a certain type of electronics. High recyclability indicates better eco-design. For instance, refrigerators have a recyclability range of 50.3～57.7; the refrigerator with a recyclability rating of 57.7 possesses better eco-design than the one with a rating of 50.3. In the development phase of a new product, when its structure and composition are planned, we can use this created measuring method to determine whether its design is more environmentally sound than before.

By measuring the recyclability, at least two phases in the life cycle of product can tremendously promote both qualitative and quantitative eco-design. In the initial manufacturing phase, a producer can employ eco-design to quantitatively choose more recyclable materials such as carbon nanotubes and lead-free solder; the packaging can also be designed for efficient dismantling and recycling. In EoL phase, the producer's eco-design rating can be scientifically determined, making a much more convincing argument for shifting the financial responsibility from producer to recycler.

4.5.3 Statistical entropy analysis along material flow

The approach is based on a comprehensive MFA and Shannon's Statistical Entropy Function that is transformed by a three-step procedure. Statistical entropy analysis was firstly proposed by Rechberger and Brunner to measure the incineration process of solid waste. The result is a new function that can be applied to any defined system with known mass flows and substance concentrations. In combination with materials balances, the method yields quantitatively the Relative Statistical Entropy (RSE) and the Substance Concentrating Efficiency (SCE) of a given system.

Until now, the method has been widely applied for the fate of carbon in waste incineration, the Chinese copper cycle, sewage sludge treatment options, waste water treatment plants, regional nitrogen budgets, lead smelting, phosphorus use, a food-based bioethanol system, and aluminum resource efficiency. Here aluminum resource efficiency was adopted to demonstrate the use of statistical entropy analysis.

Aluminum has a high recycling rate among all the metal elements, contributing to both energy saving and emissions reduction. The production of one ton aluminum through recycling secondary aluminum resource consumes only 2 800 kWh of electricity and generates only 600 kg of carbon dioxide, leading to 95% electrical energy consumption and GHGs reduction. Based on the bottom-up life cycle assessment method, the environmental effects of bauxite, aluminum oxide and electrolytic aluminum were calculated in China, accounting for approximately 1.4%, 8% and 90.6% of the overall environmental burden, respectively. Furthermore, besides energy savings and emissions reductions, aluminum recycling can create new employment opportunities, generate more economic revenues, reduce the overall solid wastes to the local landfills, and promote sectoral cooperation.

In recent years, the COVID-19 pandemic seriously influenced industrial supply chains, but at the same time created a new opportunity for countries to prepare their own recovery plans toward sustainable production and consumption. Also, many countries have prepared their carbon neutrality targets to respond climate change. For instance, China is committed to reach carbon peak by 2030 and achieve carbon neutrality by 2060. Under such a circumstance, improving resource efficiency is conducive to decoupling economic growth from greenhouse gas emissions. It is therefore essential to promote aluminum recycling, improve aluminum resource efficiency and mitigate corresponding emissions from aluminum production.

1. Method

The life cycle of aluminum in the social and economic system is divided into four stages: production, fabrication & manufacturing, use and waste management (see Figure 4 - 16). Aluminum production is composed of bauxite mining, alumina refining, aluminum electrolysis and other production links. Such material flows mainly include: ① The flows between each link and its application fields; ② The flows between production, waste recycling and the natural environment. Production and consumption stocks, as well as substances discharged into the environment, are considered losses.

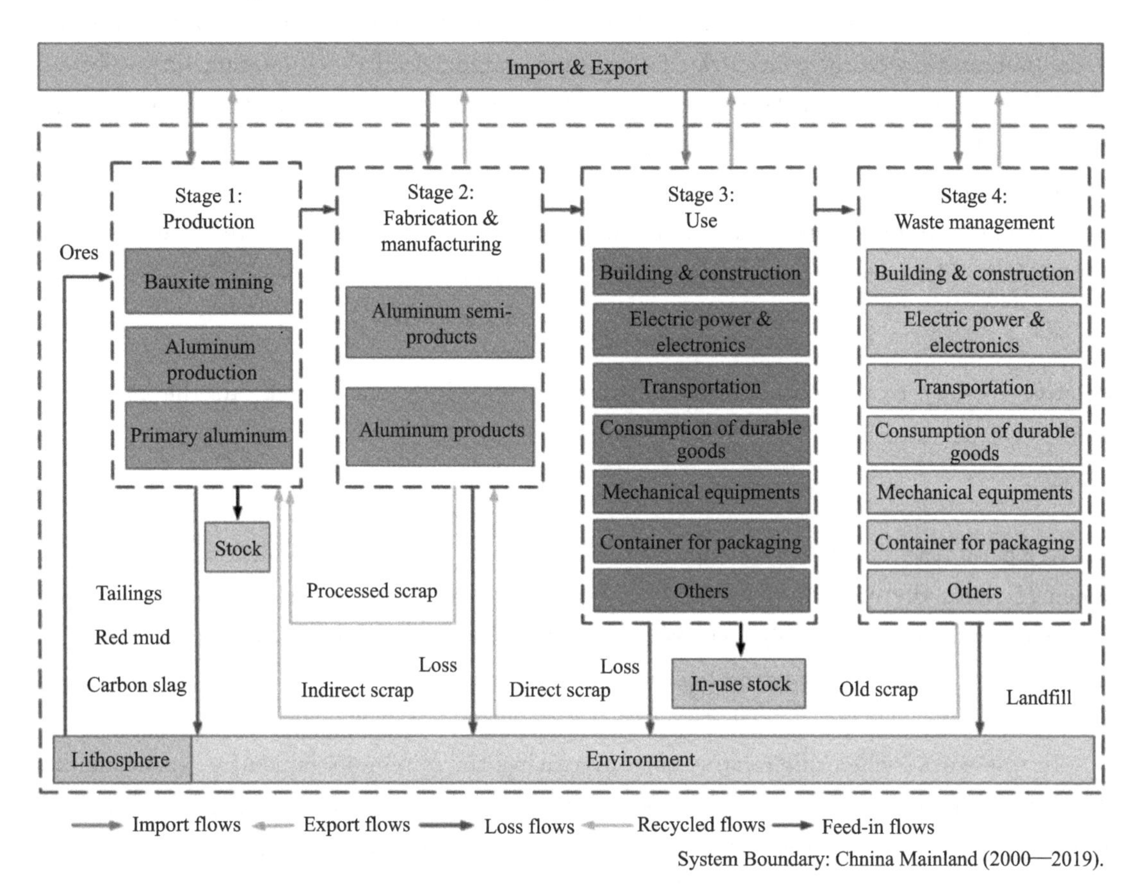

Figure 4 - 16 STAF framework applied to the aluminum life cycle

The entropy analysis method refers to the application of entropy theory in material flow analysis. This method calculates the relative entropy at a certain node according to material flows and evaluates the degree of material concentration and dissipation based on relative entropy. The results of entropy analysis can help prepare policies for improving aluminum efficiency. The following formulae are used for entropy analysis:

$$X_i = M_i c_i$$

$$m_i = \frac{M_i}{\sum_{i=1}^{k} X_i} \tag{4-12}$$

$$H(c_i, m_i) = -\sum_{i=1}^{k} m_i c_i \log_2(c_i)$$

where M_i is the weight of the ith aluminum flow, with a unit of megaton (Mt); c_i is the aluminum concentration of the ith aluminum flow; X_i is the aluminum weight of the ith aluminum flow, with a unit of megaton (Mt); m_i is the ratio of the aluminum weight of the ith aluminum flow to the concentrated aluminum weight of all the aluminum flows (totally there are k aluminum flows); H represents the entropy of the investigated system. When the material content in a regional system is the same as that in the natural environment, its entropy reaches an extreme value. C_{EC} is the aluminum concentration rate in the natural environment. Aluminum is the most abundant metallic element in the earth's crust (8.23% by weight). The formula for calculating this entropy is

$$H_{max} = \log_2\left(\frac{1}{C_{EC}}\right) \tag{4-13}$$

Relative entropy is the ratio of the entropy calculated by the system to the maximum entropy. This value can be calculated by using Eq. (4-14):

$$RE = \frac{H}{H_{max}} \tag{4-14}$$

where RE represents relative entropy at each node and H represents information entropy. When H takes the maximum value, it is represented as H_{max}, indicating that the chaotic degree of this system is the highest. When H takes the minimum value of 0, it indicates that the degree of system chaos is the lowest. In other cases, the value of H ranges between 0 and H_{max}.

In this work, when this relative entropy reaches the minimum value of 0, it means that the aluminum resources are presented at the maximum concentration, namely pure aluminum. When the relative entropy is equal to 1, it means that the entropy of this system reaches its peak, indicating that the distribution of aluminum concentration in this system is the same as that in the crust. Based on the relative entropy change, the concentration and dissipation degree of aluminum can be quantitatively evaluated. Because it adds an assessment dimension which has been missed in the traditional MFA studies, it helps further evaluate the overall resource efficiency related to materials and their industrial systems.

According to the characteristics of China's aluminum cycle and the entropy analysis method, a life cycle-based entropy analysis method is proposed. Figure 4 - 16 shows the aluminum life cycle. This life cycle is divided into four stages: P_1, P_2, P_3 and P_4, referring to raw material production, fabrication & manufacturing, use and waste management stages, respectively. There are nine aluminum flows in these four stages, which are represented by F_1, F_2, …, F_9. Figure 4 - 16 illustrates a life cycle-based entropy analysis model. Figure 4 - 17 illustrates a line chart of relative entropy, in which five nodes refer to aluminum material losses generated from different life cycle stages, namely node 1 to node 5. Then, the entropy value at each node is calculated. A higher relative entropy value means that more aluminum is dissipated at this node; while a smaller value means more aluminum is concentrated at this node. Changes in aluminum concentration and dissipation in different life cycle stages are then further analyzed.

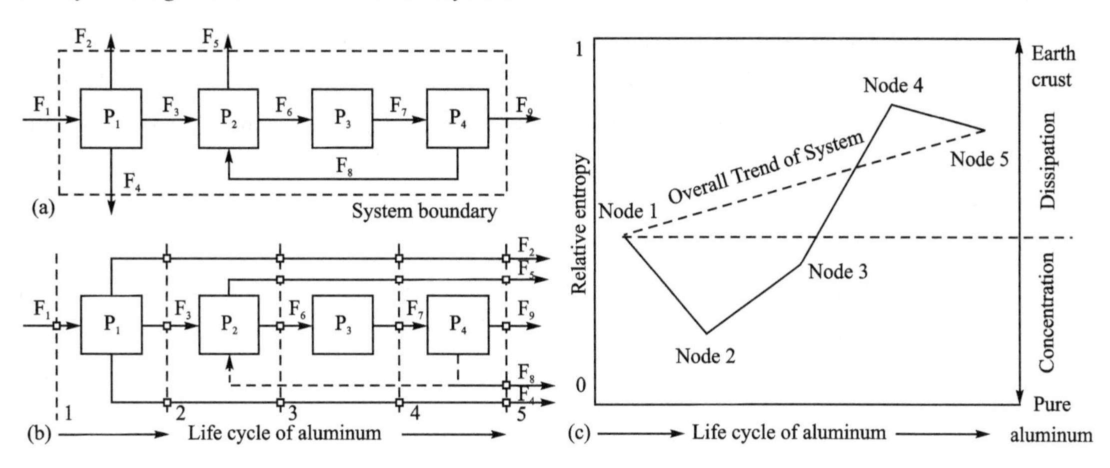

Figure 4 - 17 Life cycle-based entropy analysis for aluminum in China

It is critical to investigate future material cycles and assess their environmental impacts so that sustainable resource management can be achieved. This work employs a life cycle-based entropy analysis method to investigate the characteristics of aluminum flows. A specific aluminum entropy analysis chart is prepared and illustrated in Figure 4 - 18. The four aluminum life cycle stages are divided into five nodes for entropy analysis. There are five flows at node 1: domestic recycled aluminum, imported scrap, imported bauxite, domestic bauxite and imported primary aluminum. There are four flows at node 2: tailings,

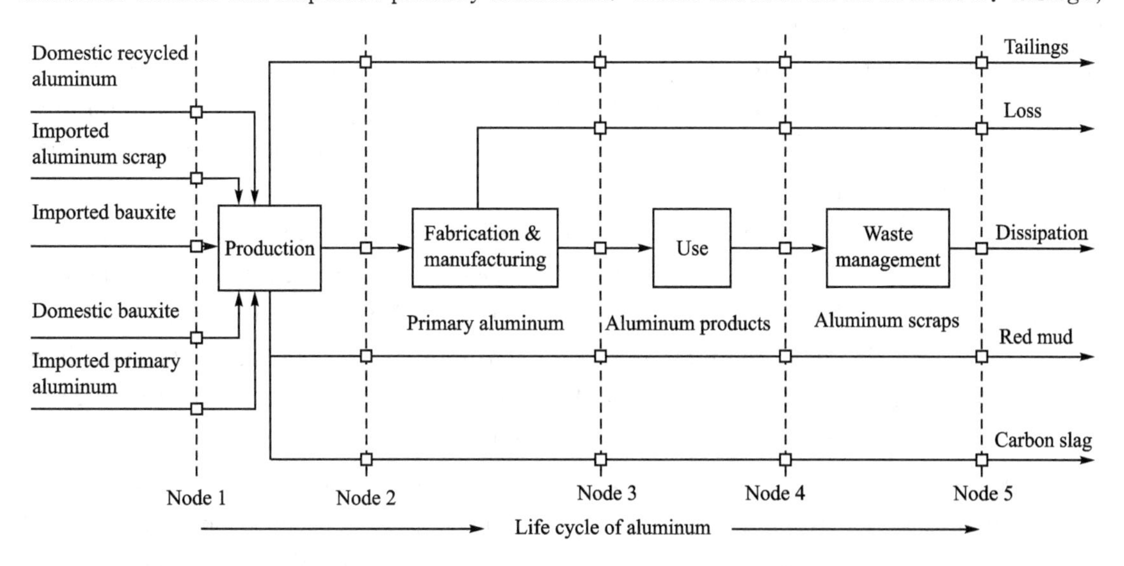

Figure 4 - 18 Life cycle-based entropy analysis diagram for China's aluminum resources

consumption of primary aluminum, red mud and carbon slag. There are five flows at node 3: tailings, loss, domestic products, red mud and carbon slag. There are five flows at node 4: tailings, loss, domestic scrap, red mud and carbon slag. Finally, there are five flows at node 5: tailings, loss, red mud, carbon slag and aluminum dissipation in the recycling process. The entropy values at five nodes are then calculated.

2. Life cycle-based aluminum flows

Figure 4 – 19 illustrates life cycle-based aluminum flows in China for years 2000, 2005, 2010, 2015 and 2019, covering production, fabrication & manufacture, use and waste management stages. In 2000, China's domestic bauxite production reached 2. 54 Mt, while 0. 11 Mt bauxite were imported, indicating that the external dependence rate was 4%. In 2019, China's domestic bauxite production reached 22. 88 Mt, while 26. 70 Mt bauxite were imported, indicating that the external dependence rate increased to 56%. Guinea, Australia, and Indonesia were major bauxite exporting countries to China, accounting for about 96% of the total bauxite import in 2019. In 2000, 0. 01 Mt alumina were exported, and the domestic alumina production reached 2. 29 Mt. In 2019, 0. 87 Mt alumina were imported, 0. 15 Mt alumina were exported and the domestic alumina production reached 38. 56 Mt. Such great changes reflect that China has experienced dynamic trade structure adjustments on both bauxite and alumina. In terms of alumina production, energy consumption is the main factor to increase its production cost, which means that alumina import can contribute to the reduction of domestic energy input and bauxite consumption. In 2019, China's primary aluminum production reached 35. 11 Mt, accounting for approximately half of the global total output, mainly supplying domestic consumption and international export markets. In 2000, the number of tailings, red mud and carbon slag generated from China's aluminum production was only 0. 33 Mt. Such a figure jumped to 7. 56 Mt in 2019, implying 21 times increase. This finding also indicates that serious environmental emissions occurred, leading to great challenges to local ecosystem and public health. The environmental loss of aluminum production accounted for 46% of the total loss throughout its entire life cycle in 2000, while such a figure increased to 69% in 2019. Consequently, it is necessary to improve the comprehensive utilization rate of tailings, red mud, and carbon slag so that the environmental risks can be reduced.

In 2000, the total aluminum consumption reached 4. 18 Mt, mainly in sectors of building & construction (1. 67 Mt), electric power and electronics (0. 44 Mt), transportation (0. 48 Mt), durable goods (0. 27 Mt), equipment manufacturing (0. 65 Mt), containers (0. 22 Mt) and others (0. 45 Mt). In 2019, the total aluminum consumption reached 32. 51 Mt, mainly in sectors of building & construction (10. 84 Mt), electric power and electronics (5. 80 Mt), transportation (7. 35 Mt), durable goods (1. 93 Mt), equipment manufacturing (2. 32 Mt), containers (3. 10 Mt) and others (1. 16 Mt). The top three aluminum consumption sectors include building & construction (33% of the total consumption), transportation (23%) and electric power and electronics (18%), accounting for 74% of the total consumption. Another fact is that many aluminum-containing products will enter the ends of their life

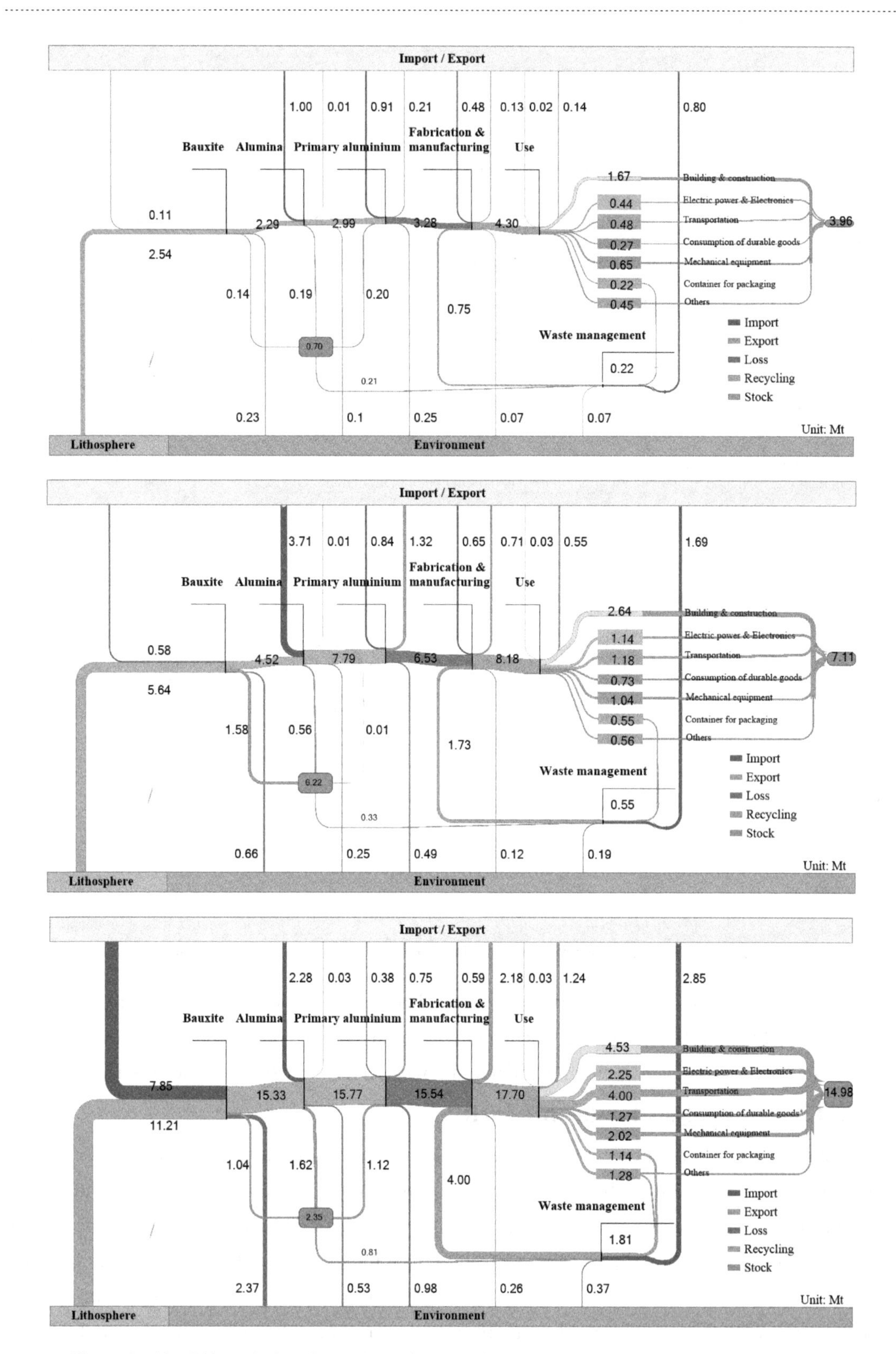

Figure 4 - 19 Life cycle-based aluminum flows in China for 2000, 2005, 2010, 2015 and 2019

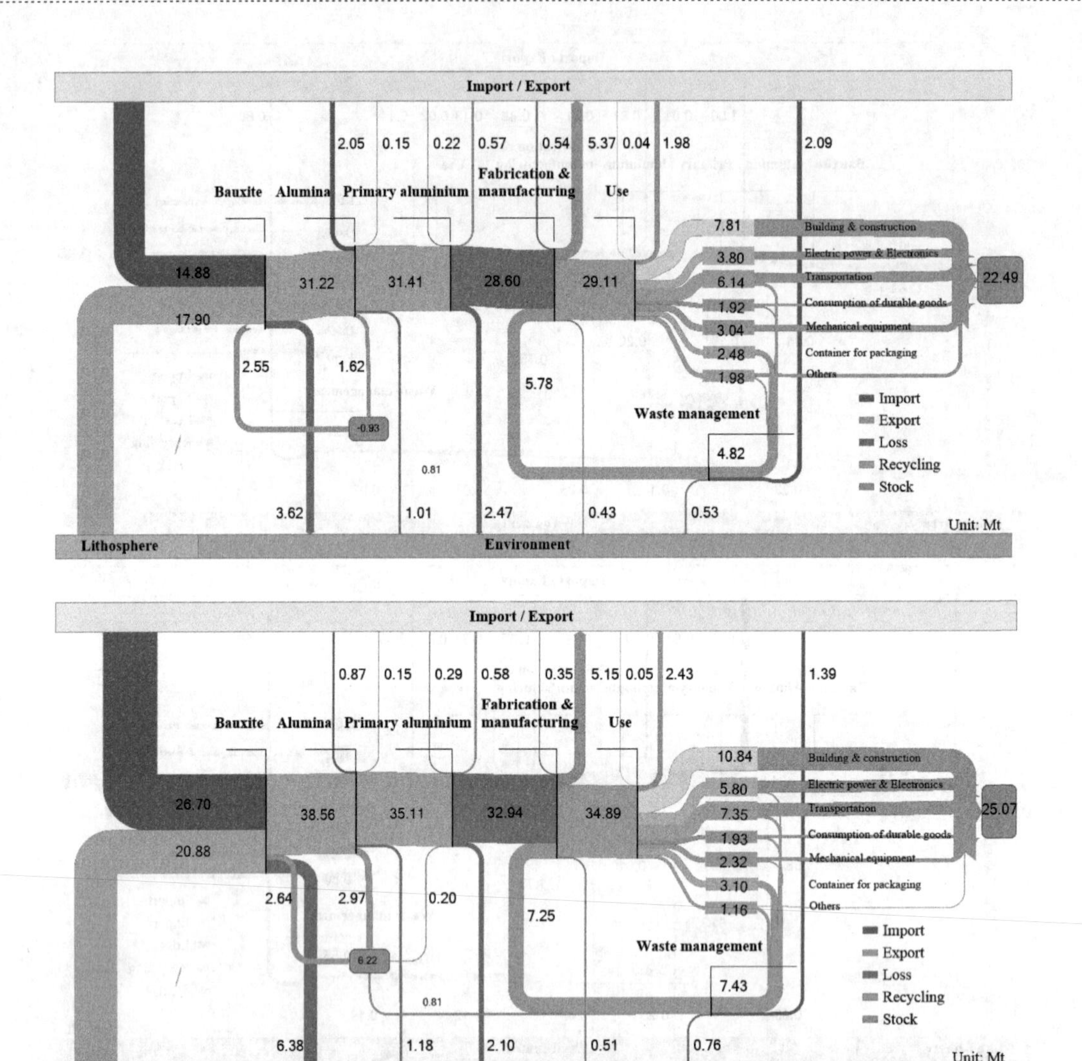

Figure 4 - 19 Life cycle-based aluminum flows in China for 2000, 2005, 2010, 2015 and 2019 (continued)

cycles. In 2019, domestic aluminum scraps reached 7.43 Mt and imported scraps reached 1.39 Mt, leading to that the self-sufficiency rate of scrap reached 84%. Technological development contributed to recycling efficiency, with clear energy and resource savings.

3. Entropy analysis for the life cycle

Based on five years intervals, Figure 4 - 20 illustrates relative entropy values at five nodes in China's aluminum life cycle for years of 2000, 2005, 2010, 2015 and 2019, reflecting the aluminum concentration and dissipation changes in different stages of its life cycle. According to the entropy analysis method, the four stages of life cycle are divided into five nodes, and there are several flows at each node. As the aluminum concentration (CEC) in the crust is 8.23% by weight, the maximum entropy (H_{max}) is calculated to be 3.602 9, and the statistical entropy (H) of each flow and the relative entropy (RE) of each node are calculated based on the mass and aluminum content of each flow.

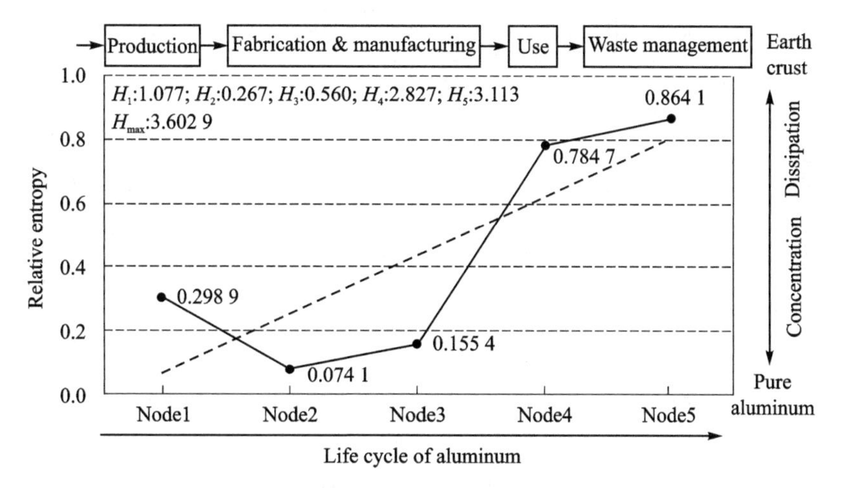

(a) The relative entropy for China's aluminum life cycle in 2000

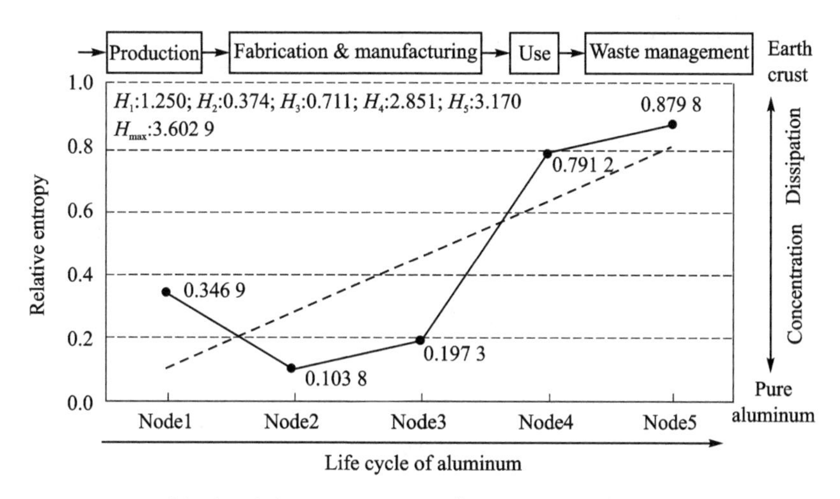

(b) The relative entropy for China's aluminum life cycle in 2005

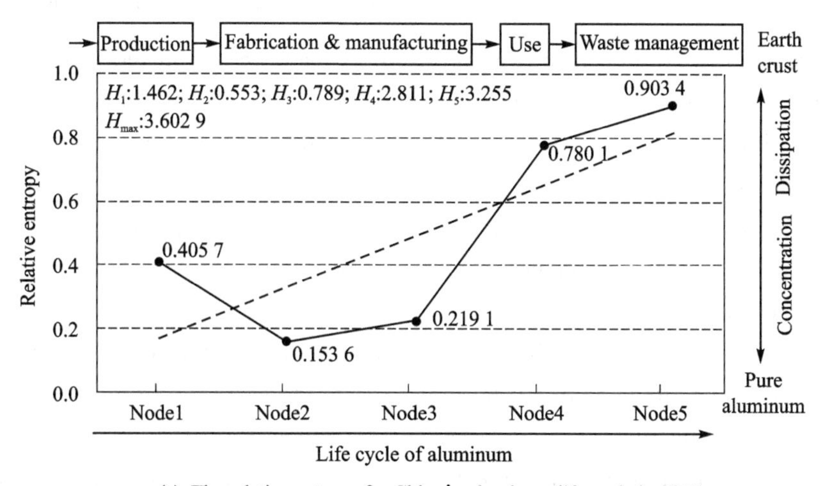

(c) The relative entropy for China's aluminum life cycle in 2010

Figure 4 - 20 Relative entropy values for aluminum life cycle in China for 2000, 2005, 2010, 2015 and 2019

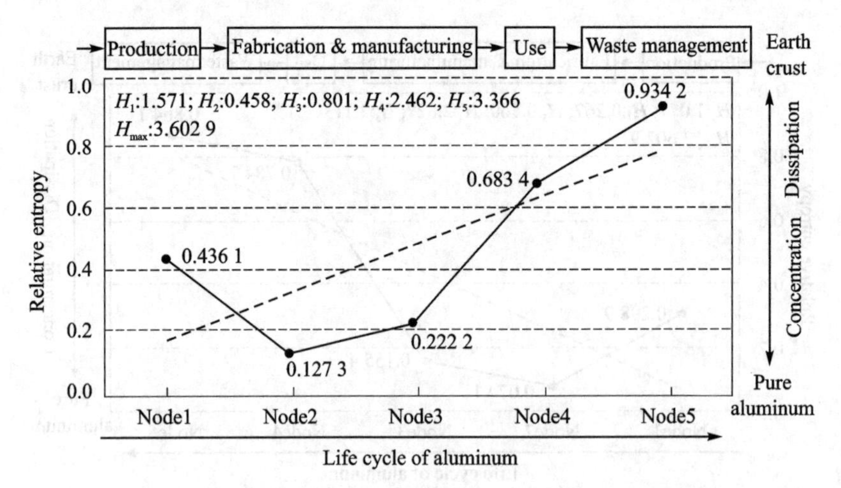

(d) The relative entropy for China's aluminum life cycle in 2015

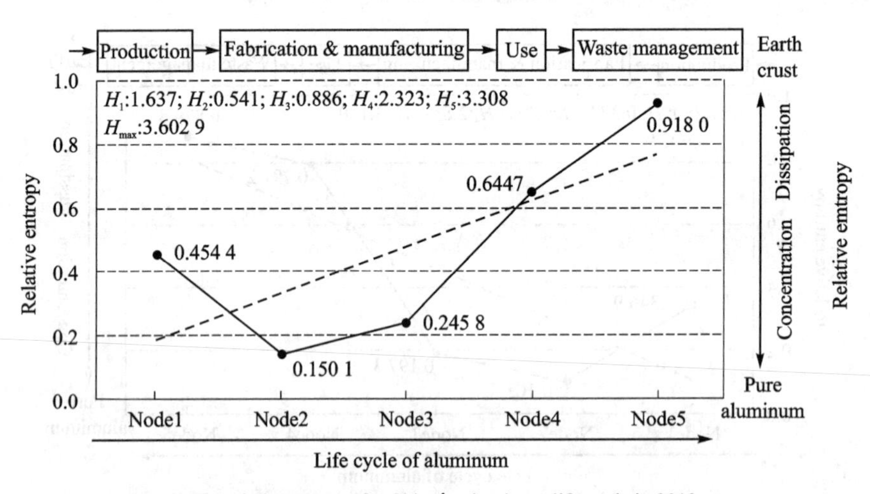

(e) The relative entropy for China's aluminum life cycle in 2019

Figure 4 - 20 Relative entropy values for aluminum life cycle in China for 2000, 2005, 2010, 2015 and 2019 (continued)

There are 5 flows at node 1, including imported bauxite, imported primary aluminum, imported aluminum scrap, domestic bauxite and domestic recycled aluminum. The relative entropy of aluminum at node 1 was 0. 454 4 in 2019, while such a value was 0. 298 9 in 2000, 0. 346 9 in 2005, 0. 405 7 in 2010 and 0. 436 1 in 2015, respectively. Such a figure reached its highest value in 2019, mainly due to the import dependence on bauxite from the production stage. The bauxite import dependence in China increased from 4% in 2000 to 56% in 2019, with an increase rate of 51%.

Nodes 2 and 3 represent the fabrication & manufacturing stages of aluminum products. At node 2, there are 4 flows, namely tailings, red mud, carbon slag and domestic primary aluminum. The entropy of node 2 in each year is lower than that of node 1. When the entropy of this node approaches 0, it means that aluminum in the stage is close to the state of pure aluminum. The relative entropy at node 2 was 0. 150 1 in 2019, while such a value was 0. 074 1 in 2000, 0. 103 8 in 2005, 0. 153 6 in 2010 and 0. 127 3 in 2015, respectively. As the production of primary aluminum increased from 2. 99 Mt in 2000 to 35. 11 Mt in 2019, environmental damages such as tailings, red mud and carbon slag also increased significantly from 0. 33 Mt in 2000 to 7. 56 Mt in 2019. There are 5 flows at node 3, including tailings, red mud, carbon slag, processing loss and domestic aluminum products. The relative entropy at node 3 was 0. 245 8 in 2019, significantly higher than that in 2000 (0. 155 4). This value at node 3 increased continuously from 2000 to 2019. With the increase of aluminum output, the processing loss also increased from 0. 25 Mt in 2000 to 2. 1 Mt in 2019, with an increase of 7. 4 times. There are 5 flows at node 4, including tailings, red mud, carbon slag, processing loss and domestic recycled aluminum. The relative entropy at node 4 was 0. 644 7 in 2019, while such a value was 0. 784 7 in 2000. The decrease of relative entropy at node 4 is mainly due to the increase of domestic recycled aluminum. Such a figure increased from 0. 75 Mt in 2000 to 7. 25 Mt in 2019, contributing to the increased aluminum concentration along its life cycle. There are 5 flows at node 5, including tailings, red mud, slag, processing loss, dissipation. The relative entropy at node 5 significantly increased from 0. 864 1 in 2000 to 0. 918 0 in 2019, indicating that there is still significant environmental damage along the whole life cycle. The accumulated environmental pollutants mainly include tailings, red mud, slag, processing loss, dissipation. Therefore, it is necessary to find an effective pathway to reduce the relative entropy along the aluminum life cycle and promote the aluminum concentration in its life cycle so that the utilization efficiency of resources in China can be improved.

4.6 Further reading

[1] Rechberger H, Brunner P H. A new, entropy based method to support waste and resource management decisions[J]. Environmental Science & Technology, 2002, 36 (4): 809-816.

[2] Huisman J, Stevels A, Stobbe I. Eco-efficiency considerations on the end-of-life of consumer electronic products[J]. Electronics Packaging Manufacturing, IEEE Transactions on 2004, 27 (1): 9-25.

[3] Kaufman S, Krishnan N, Kwon E, et al. Examination of the fate of carbon in waste management systems through statistical entropy and life cycle analysis[J]. Environmental Science & Technology, 2008, 42 (22): 8558-856.

[4] Hellweg S, Milà i Canals L. Emerging approaches, challenges and opportunities in life cycle assessment[J]. Science, 2014, 344 (6188): 1109-1113.

[5] Richa K, Babbitt C W, Gaustad G. Eco-efficiency analysis of a lithium-ion battery waste hierarchy inspired by circular economy[J]. Journal of Industrial Ecology, 2017, 21 (3): 715-730.

[6] Boardman A E, Greenberg D H, Vining A R, et al. Cost-Benefit Analysis: Concepts and Practice[M]. Cambridge: Cambridge University Press, 2018.

[7] Zeng X, Mathews J A, Li J. Urban mining of e-waste is becoming more cost-effective

than virgin mining[J]. Environmental Science & Technology, 2018, 52 (8): 4835-4841.

[8] Roithner C, Cencic O, Rechberger H. Product design and recyclability: How statistical entropy can form a bridge between these concepts—A case study of a smartphone[J]. Journal of Cleaner Production, 2022, 331: 129971.

[9] Auras R A, S E S. Life Cycle of Sustainable Packaging: From Design to End-of-Life [M]. New York: John Wiley & Sons, 2022.

4.7 Exercises

1. Estimate an LCA of a simple product, such as tooth brush, shopping bag, and so on, to find priorities for sustainability and make fair comparisons between alternatives.

2. Apply LCC analysis to your mobile phone regarding different ways of disposal when you want to replace it with a new one.

3. What are the seven approaches of eco-efficiency? (Hint: REDUCES)

4. What are the physical implications of statistical entropy analysis along material flow? Please use one case to illustrate.

5. A throw-away society and a lack of waste management infrastructure have created an unsustainable situation for packaging worldwide, becoming excruciating in low- and middle-income economies where the collection and management of packaging are scarce or insignificant. Please design one sustainable solution for packaging system using MFA and emerging methods.

Chapter 5 The Way Forward

MFA is an increasing tool (or even philosophy) and crucial to unlock some interdisciplinary problems from natural science to social science. In particular, the contemporary problems and challenges we are confronting since 2010s are much more complicated. To examine them, MFA must be also growing from depth to width. This chapter will address the evolution of MFA in the related dimensions, explore some methods related to emerging technologies and data mining, and propose the new fields for the future decades.

5.1 Evolution of material flow analysis

5.1.1 Materials

MFA originated from the focus of waste recycling and management, which sought to improve waste processing system. While this tool can imply the secret of urban system through an urban metabolism, the material of MFA has been gradually extended to other substances. The expanded scope of concerned materials covers solid waste, metal, plastic, fiber, construction minerals, composite material, agricultural products, and water. Through a bibliometric analysis, an increasing number of articles related to MFA have been published (Figure 5 - 1). Until now, water is the most concerned material as around 900 articles are nowadays published annually. Metal and plastic rank the second and third as the most concerned material. Metal and plastic are closely related to resource and environmental security, respectively, which was recognized as the grand challenges in the 21st century.

Fiber and composite material are also concerned in recent years, especially from the manufacturing industry. Agricultural product and construction mineral are the emerging materials using the MFA. The use of MFA is evolving from the heavy industry to the light industry and agriculture processing.

5.1.2 Boundary

Strictly speaking, the stocks and flows of materials on and around earth are combinations of the stocks and flows related to both natural and anthropogenically related flows. Such a perspective brings into view flows and stocks that enter and leave the surface of our planet from the interior (volcanic eruptions, subduction of tectonic plates, etc.), the oceans (chemicals in rivers and aerosols), the atmosphere (emissions from trees, surface deposition, and industry), and earth orbit (captured interplanetary dust, deorbital flow). Rauch and Pacyna generated such MFAs for the global silver, aluminum, chromium,

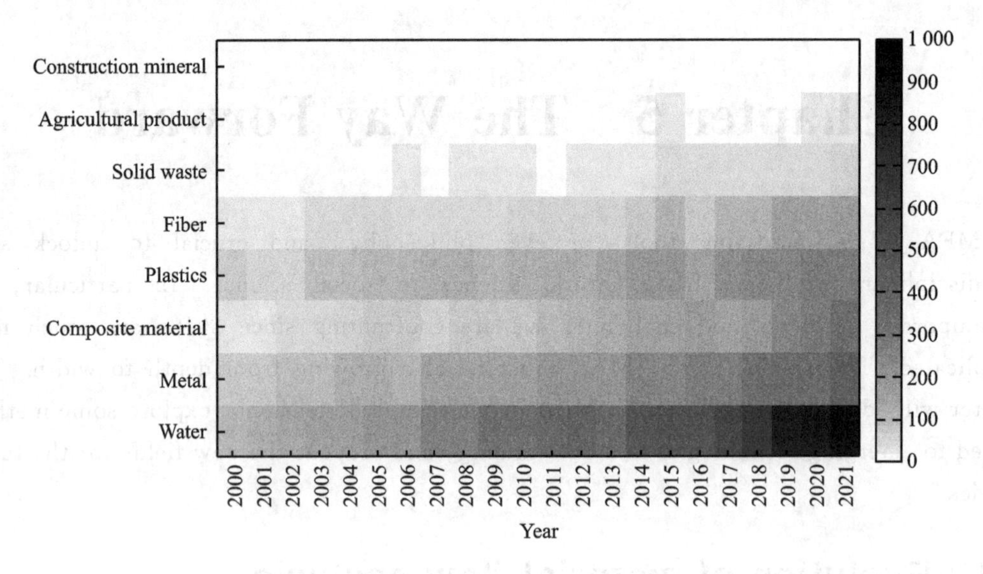

Note: search "material flow analysis" as abstract and "material (i. e. water, metal, ..., construction mineral)" as abstract from Web of Science in all the database (https://www.webofscience.com/wos/alldb/basic-search).

Figure 5 - 1 Publication number of different categories of materials among material flow analysis in the year of 2000—2021

copper, iron, nickel, lead, and zinc cycles; they found that anthropogenic activity has significantly perturbed earth's natural biogeochemical cycles, as shown by the copper example in Figure 5 - 2. In doing so they demonstrated that humans today mobilize about half the metal mass of these global elemental metal cycles. Such a broad perspective could become increasingly relevant should wide-scale mining of the seafloor or of asteroids occur in the future.

Figure 5 - 2 Potential boundary of MFA

5.1.3 Environmental effect

Material flow and associated substance release cause the environmental effect in a direct or indirect way. Environmental effect has been well addressed in many textbooks. According to Einstein's theory of special relativity, mass and energy are related to each other. Energy flow analysis has been well initialed to optimize the energy consumption and saving. Therefore, energy flow and carbon emission are the popular environmental effects. Figure 5 - 3 illustrates energy flow and carbon emission along a closed-loop supply chain and their mathematic relationships. Energy saving and emission reduction could be scientifically examined in this approach and thus governed for the government.

By this philosophy, the acquisition, processing, consumption, and recycling of materials all involve the use of energy and usually the carbon emissions to the atmosphere, some 8% of the global total. MFA analyses of the underlying materials cycles can provide

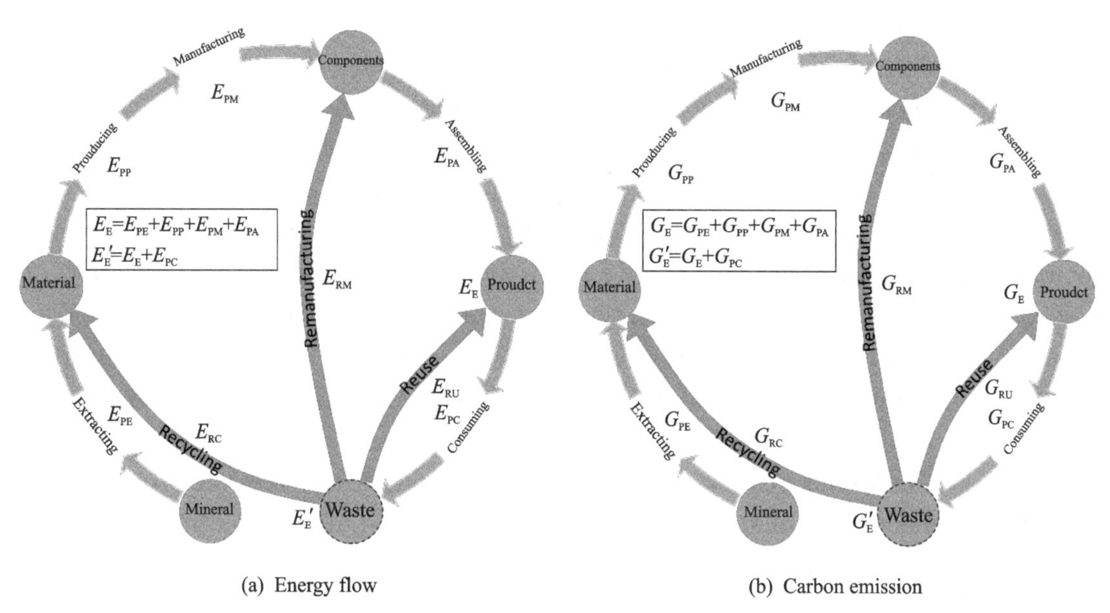

(a) Energy flow　　(b) Carbon emission

E_{PE}—the energy input in extracting from mineral to material; E_{PP}—the energy input in material producing; E_{PM}—the energy input in manufacturing to components; E_{PA}—the energy input in assembling from components to product; E_E—the embodied energy in product; E_{PC}—the energy input in consuming from product to waste; E_{RU}—the energy input in reuse from waste to product; E_{RM}—the energy input in remanufacturing from waste to components; E_{RC}—the energy input in recycling from waste to materials; G_{PE}—the carbon emission in extracting from mineral to material; G_{PP}—the carbon emission in material producing; G_{PM}—the carbon emission in manufacturing to components; G_{PA}—the carbon emission in assembling from components to product; G_E—the carbon emission embodied in product; G_{PC}—the carbon emission in consuming from product to waste; G_{RU}—the carbon emission in reuse from waste to product; G_{RM}—the carbon emission in remanufacturing from waste to components; G_{RC}—the carbon emission in recycling from waste to materials.

Figure 5 - 3　Energy flow and carbon emission along a closed-loop supply chain and their mathematic relationships

the basis for detailed analyses of carbon emissions. Here is an example of this analytical idea, Liu et al. constructed a global aluminum cycle (for the year 2009) and adopted the results as the basis to measure carbon emissions to 2050 under different assumptions of emissions control and rates of industrial growth. Similar works for other metals and materials can easily follow this approach provided the underlying MFAs have been constructed.

5.1.4　Policy analysis

In terms of policy, MFA can be used for early recognition, priority setting, to analyze and improve the effectiveness of measures and to design efficient material management strategies in view of sustainability. MFAs have been used in corporate policy analysis since 2000. In a path-breaking example, the Toyota Motor Company in 2003 presented a corporate MFA diagram entitled "Volume of Resources Input and Volume of Substances Released into the Environment in FY2002". With the MFA cycle quantified, the corporation was then able to design future aims for material use, emissions, and use of recycled materials, and follow progress year by year. Similar diagrams for other corporations have not been publicly available, but in many cases progress related to materials use and loss is summarized in corporate reports and elsewhere.

Regarding the government policy, MFA results are policy relevant, but not policy prescriptive. The role of MFA practitioners in this regard, therefore, is to generate, interpret, and communicate potentially relevant information rather than to advocate for policies. Examples of the use of MFA in this way have occurred for at least a decade, and are regarded as sufficiently reliable to serve as a basis for policy use. A few examples illustrate the utility of MFA in government policy. The flow of materials in Japan provided the justification for Japan's 3R (reduce, reuse, recycle) policies. An MFA of polybrominated diphenyl ethers in Vienna pointed to a focus on product waste and recycling plants. In the New York City harbor, MFA studies by Boehme et al. quantified flows of five different toxins and assigned them to specific industrial sectors, generally with a recommendation for remedial action. Finally, Eckelman and Chertow's MFA study of waste management flows in Oahu, Hawaii indicated several opportunities for using waste resources to substitute for imports while simultaneously reducing waste generation. A more general use of MFA for policy purposes involves critical materials determinations. Many aspects of those determinations for transboundary import and export flows, recycling performance, etc. draw from MFA analyses of potentially critical materials in national or regional policy regimes.

5.2 Emerging technologies and methods in related field

5.2.1 System analysis

System analysis is conducted for the purpose of studying a system or its parts in order to identify its objectives. It is a problem-solving technique that improves the system and ensures that all the components of the system work efficiently to accomplish their purpose. Analysis specifies what the system should do. System design involves modeling, analysis, synthesis, and optimization. System analysis is a procedure or approach that serves to determine the system's performance for a given (known) structure of this system. An example may be a typical student project with a given input data which should be made for a defined system structure. The resulting calculation data characterizes system outputs.

A system is a general set of parts, steps, or components that are connected to form a more complex whole. For example, a computer system contains processors, memory, electrical pathways, a power supply, etc. For a very different example, a business is a system made up of methods, procedures, and routines. The first step in solving a problem that involves a system is analyzing that system. This involves breaking it down into the parts that make it up, and seeing how those parts work together. Sometimes figuring out how a system works can involve turning off parts of the system and seeing what happens, or changing parts of the system and seeing what the result is. If you change what goes into a system, how does it change what comes out? Basically, systems analysis involves techniques that allow you to understand how a system works.

The world is complex and full of problems to solve. It's probably not surprising, therefore, that problem solving is one of the most sought-after skills. If you can break a problem apart, and come up with a solution, your skills will always be needed. One type of problem solving is called systems analysis. Systems analysis is a problem-solving method that involves looking at the wider system, breaking apart the parts, and figuring out how it works in order to achieve a particular goal. But before we get into detail about how that works, we should probably first answer the question: What is a system?

For instance with Figure 5 - 4, resource sustainability has moved over many system levels from end-of-pipe (combating pollution) to root cause (design for the environment). Attention has over time moved from end-of-pipe solutions (1) to more focus on clean production (2) recycling, slimmer consumption patterns and sustainable production (3). Ultimately the world must also address the consumption volume as a function of per capita use as well as the number of consumers, directly proportional to the size of the global population (4).

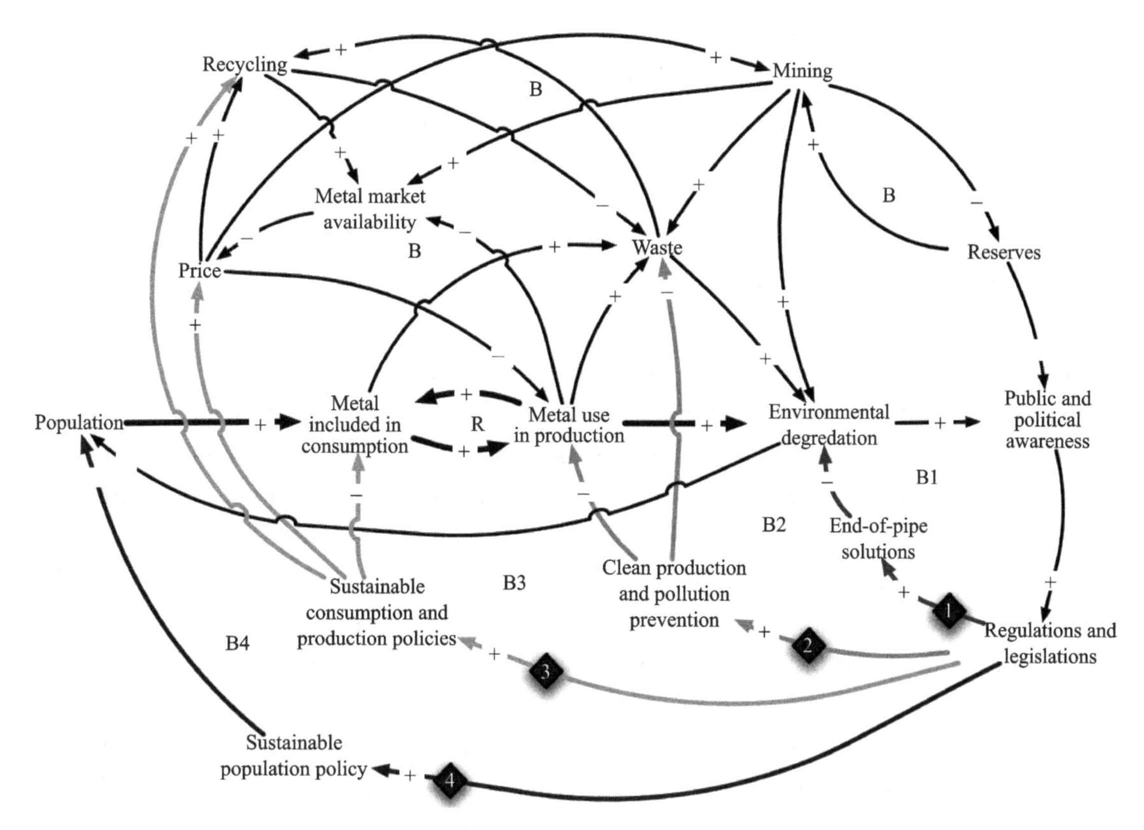

Note: B1～B4 are different balancing loops that can be introduced into the system by policies.

Figure 5 - 4 Sustainability of resource use has moved over many system levels from end-of-pipe to root cause

5.2.2 Data science

Data science continues to evolve as one of the most promising and in-demand career paths for skilled professionals. Today, successful data professionals understand that they must advance past the traditional skills of analyzing large amounts of data, data mining, and programming skills. In order to uncover useful intelligence for their organizations, data scientists must master the full spectrum of the data science life cycle and possess a level of flexibility and understanding to maximize returns at each phase of the process.

Python is an easy to learn, powerful programming language. It has efficient high-level data structures and a simple but effective approach to object-oriented programming. Python's elegant syntax and dynamic typing, together with its interpreted nature, make it an ideal language for scripting and rapid application development in many areas on most platforms. MFA driven by Python, called PyMFA, is an emerging smart tool.

1. PyMFA analysis DSL

The stand-alone applications provide an interface that allow users to create material flow systems using graphical tools. It is possible to implement such an analysis designer for the web in Javascript, but such an implementation would exceed the scope of this project. For this reason, it has been decided to design a simple domain-specific language which uses the CSV format for serialization. It is intended to be used by researchers who do not possess programming knowledge. One table in the reference[①] shows an example analysis as it could be represented in a spreadsheet calculator.

The first row of the table should define the row headers and time indices. Subsequent rows define links in the system, one link per row. Nodes are never defined explicitly; they are created implicitly the first time they are mentioned by a link definition. The same is true for stocks: they do not need to be specified and are created implicitly wherever not all material is forwarded from a given node. Links should be in order of their location in the system, meaning that links that occur later in the material flow should come later in the input file. More specifically, any link that transfers materials into a specific node must come before any link that transfers material out from that node. Every link is involved with two nodes, a source node and a target node. There are currently five types of links:

① Inflow: This kind of link does not specify any source information, and that is why cells 2～4 remain empty. Inflow links are used to provide the system with raw materials. Hence, the value stored for each time index represents a specific amount of material with a particular unit.

② Rate: A rate link simply takes the amount of material present at the source node for the given fraction and multiplies it with the transfer coefficient provided as data for each time index, forwarding the resulting amount of material to the target node. This means, that the

---

① Carol Alexandru. pymfa: A Tool for Performing Material Flow Analyses in Python 3. 2013, https://files.ifi.uzh.ch/hilty/t/Literature_by_RQs/RQ%20222/2013_Alexandru_PYMFA_Tool_for_Performing_Material_Flow_Analyses.pdf.

sum of transfer coefficients of links leaving a given node for a single material should be 1 if all material should be forwarded from the node. If the sum is smaller than 1, a stock is created. If the sum is greater than 1, a negative stock is created, which probably represents an error in the input data.

③ Fraction: Fraction links behave exactly like rate links, but they require different source and destination materials. These links are used to split a fraction from a source node into multiple fractions leaving the node. An example could be that mixed materials from recycling may be split into different fractions such as plastic and different metals.

④ Conversion: Conversion links also behave similarly to rate links, but they do not produce any stock because the factors provided as data are simply applied to convert units. This makes it possible to for example convert "pieces" to "kg".

⑤ Delay: a delay link forwards the materials according to a Weibull Distribution over time. This means that materials which flow into a node in a particular year are distributed over several years when leaving the node. The value stored in each time index represents the alpha and beta parameters of the Weibull distribution.

Rate, fraction and delay links do not need to specify target units because they should be equal to the corresponding source units. Likewise, rate and delay links do not need to specify target materials either, because they should be equal to the source materials as well. All links may specify free text in the description column.

Using the CSV format to formulate analyses has the great advantage that it is possible to create and view analyses in Microsoft Excel. Analyses could also be created programmatically from other sources, such as databases. Drawbacks include the fact that CSV is hard to read in plain text and that there exist different CSV dialects. Different formats, such as XML and JSON have been considered, but none of them can easily be created without having specific programming knowledge.

The DSL attempts to require minimal information. Because only the links in a system need to be specified, the cognitive load on the person creating an analysis remains low. Nodes and stocks do not need to be specified separately, neither is it necessary to connect the nodes and links in a particular way. One particular drawback of the DSL is that there is only one row available for all information on a given link. For the delay links, this means that for example alpha and beta parameters for the Weibull distribution have to be specified in a single cell, separated by the "|" character. Should the DSL be extended with even more complex link types, the definition of such links may become increasingly cumbersome.

In general, CSV is easy to work with for both developers and users who do not possess programming knowledge. Should a graphical editor be developed for PyMFA, it would both be possible to keep the DSL and create source files programmatically or to discard the DSL and work with a more structured format such as JSON or XML.

2. Implementation

In this section, the different components of PyMFA are presented. The core component needed to perform analyses is written in pure Python and depends only on the SciPy library[①]. A script is provided for running ad-hoc analyses without using the server, and the core components can be used as a library as well. The web server utilizes the light-weight CherryPy web framework[②] and makes use of Jinja[③] for templating. For reading and writing CSV and JSON files, Python standard library components are used. The client-side Javascript implementation depends only on the d3. js visualization library[④]. The detailed process can be found at Appendix A.

3. Visualization

The PyMFA web application allows users to view analyses online. On the analysis page, a Sankey chart is drawn to give an overview on the nodes and links of the system. The material flows during each time index are summed up and the Sankey chart hence represents the total flows of materials over all time indices. The user can now click on a link or node in the Sankey chart to drill-down and reveal more detailed information in form of a bar chart. For nodes, the amounts of different materials flowing into the node are visualized, together with the stock of each material. For links, the amounts of material transferred are drawn.

Figure 5 - 5 shows an extract of the analysis page for a particular analysis. In this example, the user has clicked on the "emissions" link. The bar chart shows two columns for each time index: the left column represents the inflow of different materials, while the right column represents stock. The colors used to differentiate materials are always the same, no

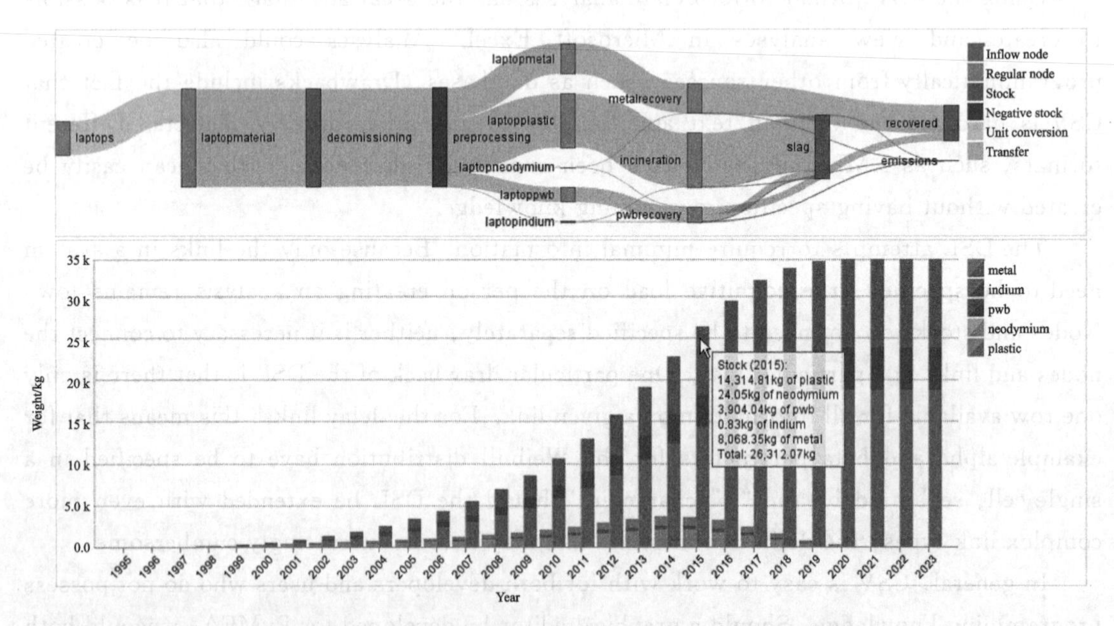

Figure 5 - 5 Viewing analysis results using the web application

① The SciPy library is part of the SciPy library stack for scientific computing in Python: https://scipy.org/.

② http://www.cherrypy.org/.

③ http://jinja.pocoo.org/docs/.

④ http://d3js.org/.

matter which node or link is visualized. This makes it easier for viewers to associate a color with a given material across visualizations. The user can hover over nodes and links in the Sankey chart and over time indices and bars in the bar chart to reveal more information. In Figure 5 - 5, the user is currently hovering over a stock column at time index 2015. The tooltip shows the exact amounts of different materials as well as the total for the given time index. Hovering over a time index label at the bottom reveals information on both inflows and stock, making it possible to view values even where the columns are too small to be targeted, as is the case for the first few years in this example.

5.3 Anthropogenic circularity for metal criticality and carbon neutrality

5.3.1 Framework and theory of anthropogenic circularity

1. Framework of anthropogenic circularity

While subsurface minerals are extracted and mined for fine metals or alloys, products are developed and manufactured (e. g. durable goods). Few compounds escape into the hydrosphere, atmosphere, and pedosphere along the entire material movement, potentially contaminating water, air, and soil. The bulk of material eventually becomes solid waste: one portion is produced as tailings, slag, home scrap, and new scrap as a result of extraction, production, or manufacture, while the other part is produced as product waste or old scrap as a result of EoL consumption. Some well-used products can be reused, some product trash can be remanufactured after disassembly, the bulk will be dismantled for recycling, and litter will be recovered for a closed-loop or open-loop as other product supply chains.

Anthropogenic circularity, driven by human activities and natural geodynamics, is reshaping the biogeochemical cycle in the anthroposphere. Natural biogeochemistry and social-economy metabolism, such as trade and logistics, make up material flow in the macroworld. From the global insight, the spatial landscape of typical metals has changed dramatically since 1960. As anthropogenic stock, geological metals minerals, for example, have been dramatically removed from subsurface to aboveground (i. e. in-use stock and waste). Currently, the twelve common metals, such as Ag, Au, Bi, Cd, Cu, Fe, Hg, In, Pd, Sb, Sn, and Zn, have a visible stock on the surface, indicating that the bulk of these metals have been extracted from the geological reserve. While anthropogenic stock of these metals in 2020 accounted for over 70% of total geological and anthropogenic resource, some geological metals like In (88%), Cd (87%), Hg (84%), Sb (82%), Sn (76%), Au (74%), and Pb (73%), are rapidly depleting. But China demonstrated a more severe situation than global circumstance, in particular on Co, Pb, Ni, and Al. Furthermore, trade plays an important function in balancing the regional landscape. China, for example, is the global manufacturing hub because it imports resources from Australia and exports the finished product to the United States. Individual mining, manufacturing, and recycling companies might pursue resource utilization and waste minimization in the microworld.

2. Metric of anthropogenic circularity

To measure anthropogenic circularity, humans just started to establish the metric or indicator system. Until now, there is still no standardized method to determine the circularity. Anthropogenic circularity could be measured at different spatial scales, ranging from macro-(national or regional) and meso-(industrial park), to micro-levels(corporation) with various metrics or indicators. For instance, recycling rate and resource efficiency are commonly employed to measure the circular economy. The recycling rate and detoxification rate are used to examine waste management. Emergy analysis is an emerging metric used for an industrial park and industry system. Recyclability is adopted to measure the recycling difficulty of solid waste and helps in product eco-design. A circular economy indicator system was proposed to measure the progress of one country or city. On this basis, the method of multiple correspondence analysis was proposed to examine 63 circularity metrics and 24 features relevant to the circular economy, such as recycling rate, resource productivity, and disposal rate. Ruiz-Pastor et al. (2022) designed a metric, CN_Con, to measure the circularity. It will help to compare quantitatively the circularity and novelty potential of different design alternatives.

5.3.2 Anthropogenic circularities against metal criticality

1. Metal criticality and its measurement

Massive metals have been seized by the rapid advancement of high technology. According to O'Connor et al. (2016), printed circuit boards contained only eleven elements in the 1980s, fifteen in the 1990s, and sixty in the 2000s. Rare metals are increasingly being used in high-tech items in contemporary culture. The emergence of criticality was prompted by shortage. Since 2010, the term "criticality" has been formalized to define the quality, state, or degree of being of the utmost importance, and it has recently sparked interest in strategic metals. Prof. T. E. Graedel at Yale University proposed and created the measurement model of metal criticality. Criticality indicates the significance of metal from two dimensions of the supply risk and the vulnerability to three dimensions of the supply risk, the environmental implications, and the vulnerability to supply restriction. The economy, geology, commerce, regulation, policy, and even politics are all variables in the three dimensions. It was calculated using three dimensions and dynamic weight.

Numerous resources have been investigated, including copper family, iron & alloy, metalloids, nuclear energy metals, zinc, tin, & lead, rare earth element, water, and specialty metals. Precious metals have the most serious environmental consequences, even when compared to some ordinary metals. Because of the dynamic nature and variance, criticality differs from nation to country. Until March 2022, over 50 articles on criticality were published, covering over 60 base metal, rare metal, and rare earth elements. Copper, indium, niobium, and a few precious metals are among the most concentrated metals. It suggests that in many countries, these metals are a top priority among all critical metals.

2. The relationship between anthropogenic circularity and metal criticality

In recent years, many studies on criticality have been translated into some applications to guide the automobile industry, the thin-film photovoltaic technologies, and transport network. China is also highlighting the critical metals issue along the whole supply chain. Tsinghua University established the measurement method of carrying capacity for critical metals based upon the closed-loop supply chain and completed over ten critical metals for China. In 2019, the National Natural Science Foundation of China initiated a major research plan related to critical metals. It is devoted to answering three key scientific questions: the multi-layer interaction of the earth and the enrichment process of critical metals, the metallogenic mechanism of critical metals, and the occurrence state and enhanced separation mechanism. The three-dimensional framework was adopted to measure the copper and aluminum in the world. The values of platinum-group metals in the three dimensions were reported to rank in the first two places, indicating a higher supply shortage risk as well as a vital and fundamental role for the Chinese economy and facilities construction.

Anthropogenic circularity can offer certain resources for industry and minimize the demand for primary (or virgin) metal through reuse, remanufacturing, recycling, and recovery. As a result of design for circularity, increasing product recyclability and longevity is also a key signal of criticality measurement. It thereby may be able to reduce metal demand supply risk and ease metal criticality. In the case of rare earth elements, for example, the production of scrap was small in the short term compared to the expanding demand. However, one efficient method of combating criticality has been established as anthropogenic circularity through recycling. A large amount of energy metals such as lithium, cobalt, indium, nickel, and gallium have been consumed in the new energy business. The associated circular business models can be employed to reduce material criticality. Hence, elevating anthropogenic circularity can bring down material criticality.

5.3.3 Anthropogenic circularities for carbon neutrality

1. Carbon neutrality and its measurement

From the dictionary, neutrality in chemical science is the condition of being chemically or electrically neutral. Carbon neutrality is defined as an environmental carbon balance sheet with zero net CO_2 equivalents (CO_{2eq}) emissions. Carbon neutrality refers to net-zero CO_2 emissions attained by balancing the emission of CO_2 with its removal so as to stop its increase in the atmosphere that causes global warming. For the aluminum industry, carbon neutrality is defined as a state where the total "in-use" CO_{2eq} saved from all products in current use, including incremental process efficiency improvements, collection, and urban mining activities, equals the CO_{2eq} extended to produce the global output of aluminum.

In the 21st century, carbon neutrality is becoming the new industry paradigm around the world. Carbon neutrality may be regarded as the fifth industrial revolution. As of February 2021, 124 countries had pledged to achieve carbon neutrality by 2050 or 2060. Confronting the urgent need to deal with climate change, the "Copenhagen Accord" and the "Paris Agreement" propelled governments to establish domestic targets to reduce carbon emissions. Its measurement is an emerging and hot issue in the world. A spurt of publication is addressing this issue. At least in China, carbon accounting is entering the industrial sectors and indicates a coming massive movement of carbon emission reduction.

2. The relationship between anthropogenic circularity and carbon neutrality

Because recycling usually uses less energy than creating new materials, increasing circularity reduces pollution while also lowering energy demand. Metal primary mining's cumulative energy requirement has a considerable impact on global warming potential (Figure 5 - 6). Higher energy demand apparently indicates higher global warming potential since all forty-three metals have a linear relationship. As a consequence, increasing circularity has the potential to reduce CO_2 emissions. For example, the primary and recycling CO_2 emissions for a one-ton copper mine are roughly 1.71 and 0.88 tons, respectively (Figure 5 - 6). Precious metals such as gold and palladium have a massive discrepancy in absolute value. But the aluminum and iron have a lower ratio so they have a comparatively high potential for CO_2 emission reduction.

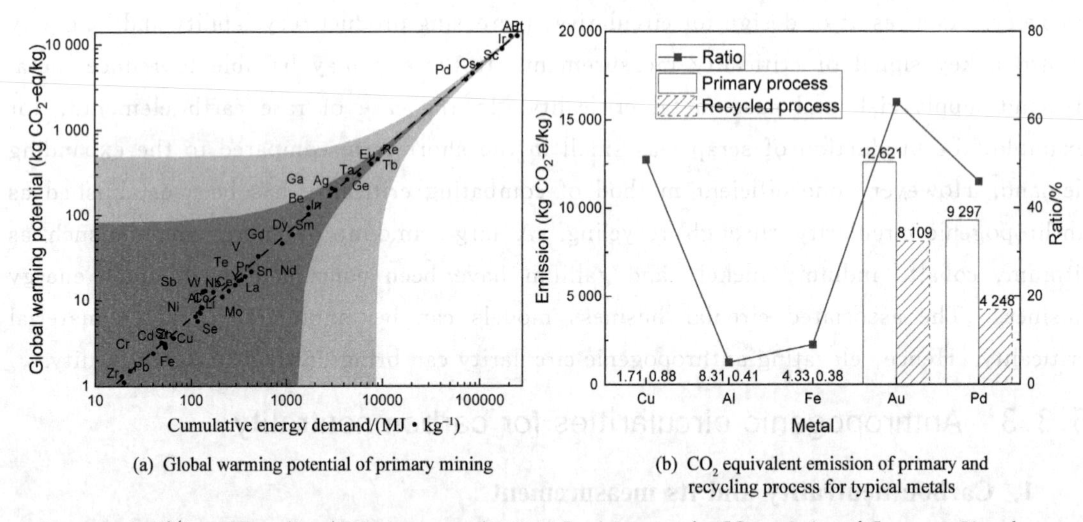

(a) Global warming potential of primary mining

(b) CO_2 equivalent emission of primary and recycling process for typical metals

Note: Ratio (%) = (Recycling / Primary process) × 100. For instance, the CO_2 emission of Cu was 1.71 and 0.88 kg per kg metal yield for primary process and recycled process, respectively. The ratio of Cu will be 51%.

Figure 5 - 6 Global warming potential of primary mining and CO_2 equivalent emission of primary and recycling process for typical metals①

Similar to metal, plastic circularity can provide environmental benefits through a circular economy. Klotz and his colleagues built the material flow system of 11 plastic types in 54 products in Switzerland in 2017 and found that 21% to 100% of the secondary material

① Data source from the references [100]~[105].

gained can substitute 11% to 17% of the total material demand. Pulling valuable metals from anthropogenic mineral makes financial sense over virgin mining of geological mineral, which could be subject to low energy input and less carbon emission in a circular economy. Critical metals, on the other hand, are necessary to support carbon-neutral decarbonization technologies such as solar and wind energy.

5.3.4 Practice and opportunities of anthropogenic circularity

1. Practice of anthropogenic circularity

At the practical level, anthropogenic circularity has been enabled. Waste recycling, clean production, circular economy, urban mining, and a zero-waste program are all common activities around the world. Material flow analysis can indicate the progress of anthropogenic circularity. Copper, for instance, is one of the most important metals to function in our high-tech industry and society. Local copper is fed after the excavation from the lithosphere and importation from other places. The fine copper is used for fabrication and manufacturing (F&M). After the consumption, the product will reach the EoL and become the product waste (or old scrap). Some waste generated in the production or F&M stages covers tailings, slag, and defective goods (or new scrap), of which the new scrap is much easier than tailings and slag in metal recycling. At the EoL stage, the old scrap like e-waste can be remanufactured and recycled, and a part as loss flows to a landfill.

Copper is mostly consumed in Europe, China, Japan, and the United States. To show copper flow practice, two time periods in the mid-1990s and the 2010s were chosen (Figure 5 - 7). Total geological mineral extraction for production was roughly 6.72 Mt from the lithosphere and 5.25 Mt from importation. Only 4.35 million tons of anthropogenic mineral was used in production, with 12% coming from old scrap and 88% from new scrap. Only 35% of demand could be met by recycling anthropogenic minerals through a circular economy. About 15.03 Mt of old scrap was buried as loss: 13%, 25%, and 15% of it went to remanufacturing, recycling, and landfilling, respectively, while the rest (47%) was not collected or handled formally. From the 1990s to the 2010s, no substantial differences were observed in Europe, Japan, or the United States. In each flow, China displayed a remarkable growth. We are still a long way from a closed-loop circularity system on a global scale.

2. Influence of anthropogenic circularity

An increase in anthropogenic circularity can help to assure metal supply security while also lowering criticality. Here using the existing method of metal criticality, I quantify the influence of anthropogenic circularity upon criticality in China. If no recycling, the criticality of copper and aluminum is 46% and 39%, respectively. With the Chinese real recycling situation, the criticality of copper and aluminum is 47% and 41%, respectively. In the case of full recycling, their criticalities reach 51% and 44%. Thus, China had 4% and 3% of the utmost increase potential of anthropogenic circularity for the criticality of copper and aluminum, respectively (Figure 5 - 8).

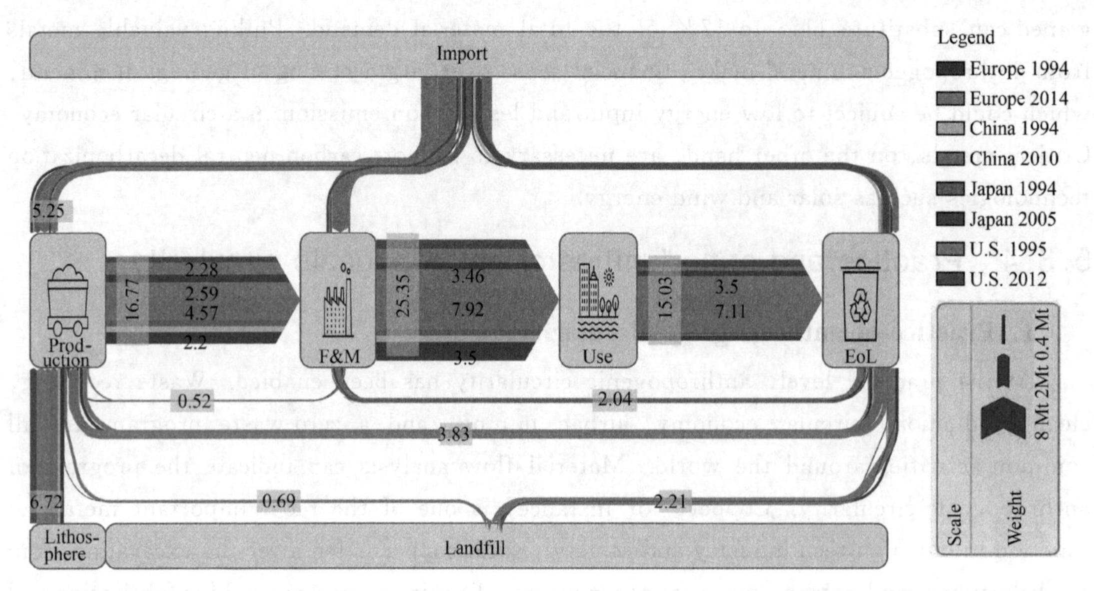

Note: stock data is not indicated here. Data source of Europe 1994 from references [107] and [108]; Europe 2014 from references [109] and [110]; China 1994 from reference[111]; China 2010 from reference[112]; Japan 1994 from reference[113]; Japan 2005 from reference[114]; and U. S. 1995 and 2012 from reference [115]~[117]. The sum of flows is also given in this figure. For instance, 6. 72 from lithosphere to production is the total flow of Europe, China, Japan, and the U. S. for the two years.

Figure 5 - 7 Substance flow diagram of copper in some countries

The formation of old scrap as product waste has the indirect potential of carbon emission reduction via an anthropogenic circularity, according to the carbon neutrality perspective. For example, in China, the three basic metals of iron, copper, and aluminum have been routinely recycled since 1949. Due to the increase in output, their total estimated emission reduction in full recycling increased exponentially from less than 1 Mt in 1949 to 1 700 Mt in 2006, with iron accounting for almost 95% of the total (Figure 5 - 8). Based on the actual recycling rate, the achieved emission reduction was expected to be between 270 Mt in 2009 and 760 Mt in 2019, with an average annual increase rate of roughly 18%.

3. Opportunities of anthropogenic circularity

Currently, the recycling rate for the majority of metals is still less than 50%, so we are far away from a closed-loop material society. As a global society, the total production of copper grew from 4. 64 Mt in 1960 to 15. 9 Mt in 2010. Meanwhile, old scrap played an increasing role in this production from 32% in 1960 to 60%~63% in 2010. Nearly half of the total zinc losses in the whole life cycle occur due to the old scrap not being collected for recycling after being

Box 5 - 1: Zero-waste city construction

A zero-waste city refers to an advanced urban development and management model that aims to promote green lifestyles, minimize the amount of waste produced, strengthen recycling programs and ensure that waste released into the environment is harmless. Construction of the pilot zero-waste city project is of great significance to promote and deepen comprehensive reform of urban solid waste management, and an important measure to realize an ecological civilization and build a beautiful China. It aims to be a replicable program that realizes the nation's zero-waste target.

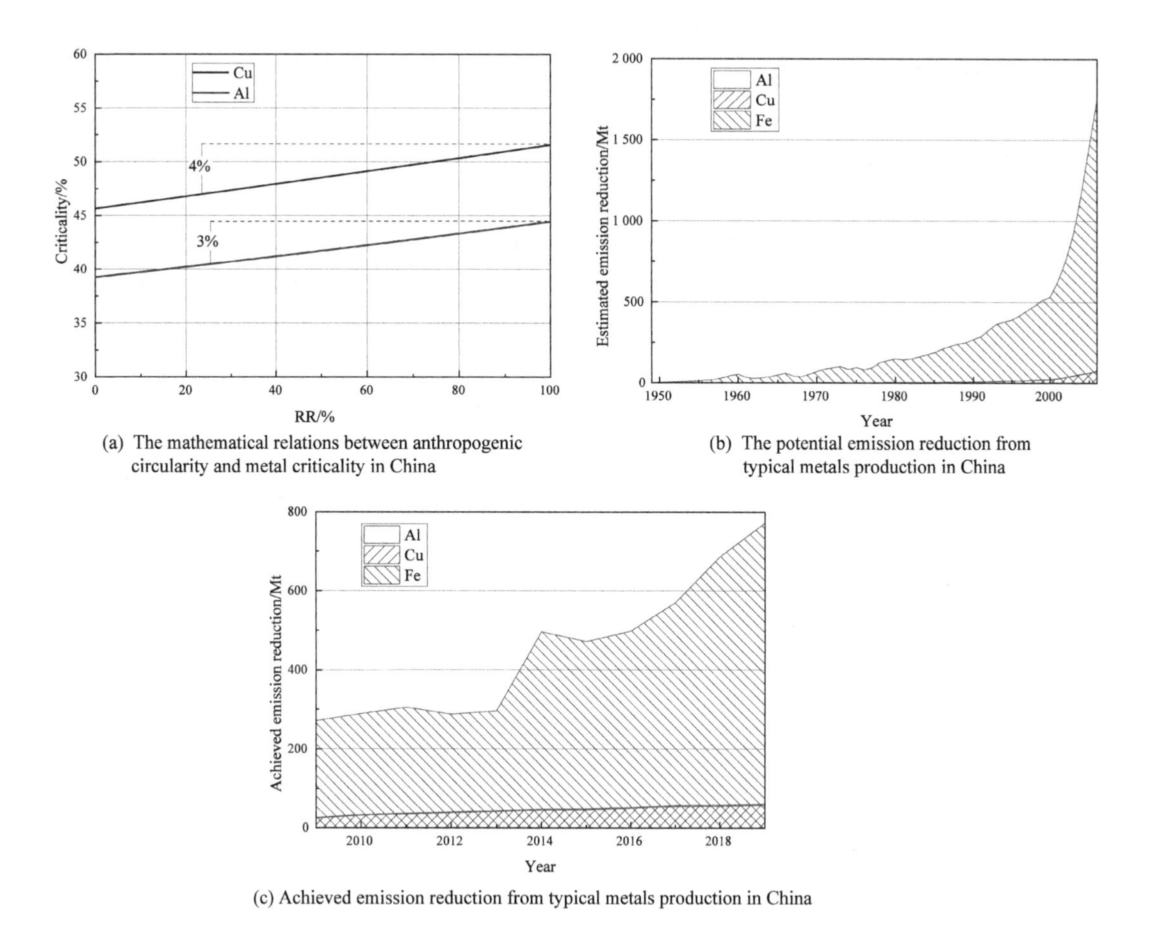

(a) The mathematical relations between anthropogenic circularity and metal criticality in China

(b) The potential emission reduction from typical metals production in China

(c) Achieved emission reduction from typical metals production in China

Figure 5 - 8 Influence of anthropogenic circularity

discarded. Hence, improving this life stage can make the largest contribution towards more circularity. From a circular economy perspective, the life cycle gap in carbon emission was 54%, 61%, and 98% for plastic bottles, rechargeable batteries, and T-shirts, respectively. Accordingly, there is a copious opportunity for further circularity improvements to achieve anthropogenic circularity and sustainability.

Circular economy and zero waste have been used to validate anthropogenic circularity in the above-mentioned practice. In recent years, the European Union and China, for example, have developed a full circular economy package or program. EoL vehicles and e-waste recycling are two examples of growing industry. The automobile sector recycles at a significantly higher rate than the e-waste industry. China launched a larger-scale trial program for a zero-waste city in 2019, with the goal of promoting all waste collection and recycling methods. After two years of success, the technology, policy, standard, and method for less generation and more collecting have been scaled up. Since 2022 China has expanded its zero-waste city program to more than 100 cities. The largest scale anthropogenic circularity in the world is embarking.

5.4 Perspectives

Despite considerable progress, challenges remain for MFA analysis. One is improving material flow statistics, especially for reuse, recycling, and remanufacturing. A second is improved identification of the materials content of multi-material products, where at present a paucity of materials-level information compromises some of the potential usefulness of national import/export statistics related to product flows. Third, most analyses of in-use stocks rely on rough estimates of product lifetimes. Can today's estimates be improved?

Finally, can MFAs be generated for such complex but widely used materials as alloys and composites, and where would the necessary data come from? Prospects for improvement in MFA accuracy and coverage appear bright, however. Missing cycles have been identified for metals (tantalum, heavy rare earth metals, thermoset polymers, etc.), and in many cases other gaps can probably be addressed, at least approximately. MFAs at one level (global, country, etc.) can also be pushed to other levels given sufficient effort and collaboration, and the usefulness of those MFAs will increase substantially as a result. Recycling and remanufacturing statistics, traditionally an MFA weak point, are likely to be improved by new and developing technology for material detection during recycling. Finally, MFA practitioners need to reach out more vigorously to the policy community in order to demonstrate the value of MFA information for policy purposes at levels from corporations to the planet. Much has been achieved through MFA development and quantification in the past few decades, much is yet to come.

5.5 Further reading

[1] Allwood J, Cullen J M. Sustainable Materials: With Both Eyes Open[M]. Cambridge: UIT Cambridge Ltd, 2012.

[2] Reck B K, Graedel T E. Challenges in metal recycling[J]. Science, 2012, 337 (6095): 690-695.

[3] Graedel T E. Material flow analysis from origin to evolution[J]. Environmental Science & Technology, 2019, 53 (21): 12188-12196.

[4] Cullen J M, Allwood J M. Mapping the global flow of aluminum: from liquid aluminum to end-use goods[J]. Environmental Science & Technology, 2013, 47 (7): 3057-3064.

[5] Allesch A, Brunner P H. Material flow analysis as a decision support tool for waste management: a literature review[J]. Journal of Industrial Ecology, 2015, 19 (5): 753-764.

[6] Cheng K L, Hsu S C, Hung C C W, et al. A hybrid material flow analysis for quantifying multilevel anthropogenic resources[J]. Journal of Industrial Ecology, 2019, 23 (6): 1456-1469.

5.6 Exercises

1. What is the typical characteristics of evolution for MFA in recent years? Why?

2. Please discuss the evolution of latest emerging technologies employed in MFA.

3. What are the three laws of anthropogenic circularity science? How about it links with MFA to foster circular economy?

4. Please outlook our circular and zero-waste society with the support of MFA.

5. Towards a socio-economic system, Japan since 2000 started to establish a sound material-cycle society[①]. Please use MFA approach to review Japan's progress and adventure.

① https://www.env.go.jp/en/recycle/smcs/index.html.

Subject Index

Subject	Page
CBA: Cost-benefit Analysis	96
ELV: End-of-life Vehicle	100
EPR: Extended Producer Responsibility	86
IPP: Integrated Product Policies	86
LCA: Life Cycle Assessment	88
LCC: Life Cycle Cost	88
LCM: Life Cycle Management	86
LCT: Life Cycle Thinking	86
MCS: Monte Carlo Simulation	35
MSW: Municipal Solid Waste	1
SLCA: Social Life Cycle Assessment	88
WEEE: Waste Electrical and Electronic Equipment	71

Appendix

Implementation process of PyMFA①

(1) Core

The core of the implementation contains the necessary code to construct and run analyses. It is defined in lib/core. py and contains the following classes:

① System(object): A container class which stores a list of timeIndices, a dictionary of nodes (using node names as keys) and a list of links of a simulation system. Once these properties are set, one can call run() on the system to start the simulation.

② Node(object): The base class for all nodes. Every node has a name, a dictionary of fractions, a reference to the systems timeIndices, and a type. This class provides a method sumStock, which, depending on the type of the node and depending on the links connected the node, calculates the stock for each fraction of the node. This function is called only once at the end of the simulation. Node also provides a function data(), which returns the node properties in a dictionary for further use. The return value of data ( ) should never be modified, as it contains references to actual node properties and does not constitute a deep copy of said properties.

③ Inflow(Node): A special kind of node which can be initialized with existing values. It is used to supply the system with material.

④ Link(Object): The base class for all links. A link has numerous properties which correspond to the cells of a single row in an analysis input file: It stores the optional free-text description provided by the user, references to the src (source) and dst (destination) nodes and it stores the source and destination materials and units as srcMaterial, srcUnit, dstMaterial and dstUnit. The amounts of material transferred are stored in an ordered dictionary values, using time indices as keys. Each node also holds a copy of the amount of material present at the source in an ordered dictionary srcValues. Furthermore, each link has a type and a reference to the system's timeIndices. Similarly to the base class for nodes, the link class also provides a function data ( ), which returns the link properties in a dictionary. There is also a helper method initDst ( ), which is called for each link at the beginning of the simulation, and which creates the necessary fraction containers inside the target node of each link. This is necessary because node definitions are implicit, and the user

① Carol Alexandru. pymfa: A Tool for Performing Material Flow Analyses in Python 3[EB/OL]. (2013-12-29). https://files. ifi. uzh. ch/hilty/t/Literature_by_RQs/RQ%20222/2013_Alexandru_PYMFA_Tool_for_Performing_Material_Flow_Analyses. pdf.

does not need to specify which fractions a node contains, as this is handled by initDst( ). Finally, the base class for links specifies two unimplemented methods, which may be implemented by its subclasses: propagate( ) implements how the material is forwarded by this link and calculateStock ( ) is a function similar to a node's sumStock ( ) in purpose and use.

⑤ Rate(Link): A kind of Link which simply forwards a given fraction of material from the source node to the destination node. Besides the parent class arguments it accepts an additional argument rates, which contains an ordered dictionary of time indices to transfer rates. When propagating values for a given time index, this kind of links simply takes the amount of material available at the source and multiplies it by the given rate for that time index, storing the result at the target node. It implements the propagate() and calculateStock() functions accordingly and extends data() to include the rates property.

⑥ Conversion(Rate): A special kind of rate which only differs in its type property and in that it does not create any stock, because conversion links are used to convert materials and units, and not forward actual material.

⑦ Fraction(Rate): Another subclass of rate which exhibits exactly the same behavior and only differs in its type property.

⑧ Weibull(Link): A kind of link which, similarly to rate, takes an additional argument parameters, which should contain a dictionary of time indices to Weibull (alpha, beta) parameter tuples. It overrides the propagate() method so that materials are forwarded with a delay. The amount of material forwarded for each time index consists of fractions of materials from several past time indices according to a Weibull distribution. The data( ) function is extended so that it includes the Weibull parameter dictionary.

The core only depends on SciPy for handling Weibull distributions and it is feasible to use the core implementation as a library for other ventures. Through its object-oriented design, it allows for the creation of new kinds of nodes and links. For example, one could implement a sink node, which discards incoming materials with just a few lines. Another plausible addition would be the creating of new delay links, which use a different algorithm to determine the delay with which materials are forwarded.

(2) Importer

The filelib/importer. py defines a class CSVImporter(object) which is responsible for reading and parsing CSV analysis source files and constructing a system instance via its load( ) method. The importer uses Python's CSV. Sniffer implementation to attempt to determine the CSV dialect of the source file, which means that it is able to understand a variety of different quoting and delimiter characters. The importer performs many sanity checks against the values contained in the CSV source file and when errors occur, it throws a CSVParserException including the row and column where the specified error occurred. This allows users to debug their source files more easily.

Like it has been mentioned earlier, analyses are defined through their links. Nodes are not specified by the user and are created implicitly whenever a link defines a particular source or target node for the first time.

(3) Exporters

The current implementation provides two exporters for serializing analyses, contained inlib/exporter. py. The CSVExporter stores the results of an analysis in CSV format, while the JSONExporter stores the entire state of the system as a JSON file. The JSON exporter is primarily used to transfer data from the web application to the client, but it could be used for other purposes as well.

(4) Web Server

The file pymfa-server. py implements a simple web server using the light-weight CherryPy web framework for Python. It uses Jinja2 to render HTML templates contained in the template folder and serves static files from the static directory. The server configuration is contained in cfg/cherrypy. ini, which also contains the user configuration. The web server offers basic functionality that enables users to upload, view and explore analyses as well as download analysis results. The server exposes the following URL scheme:

① /index: Shows the list of existing analyses. Users who are logged in are able to delete their own existing analyses from this page. The admin user can delete any analysis.

② /analysis/<name>: Serves HTML, CSS and Javascript code that allows users to view the analysis with the given name. The visualizations utilized are discussed. If the analysis with the given name does not exist, a different page is served, providing an upload form for the user to create a new analysis. The upload page also provides instructions on how to create a valid source file.

③ /analysisJSON/<name>: Serves the JSON representation of the simulation system and its results. This URL is used by the visualization components as well.

④ /source/<name>: Serves the original, unmodified source analysis file uploaded by the user.

⑤ /results/<name>: Serves the results of an analysis in CSV format.

⑥ /upload: Upload handle which is used by the upload form to submit a new source file to the server.

⑦ /login: Handle used by the login form to authenticate users and create a session for them. The handle is protected by HTTP Basic Authentication, which serves as a simple login mechanism. Note that HTTP Basic Authentication is not secure over plain HTTP. If security is a concern, HTTPS must of course be used.

⑧ /logout: Handle used by the logout button to invalidate user sessions. Returns a "401" HTTP status code which causes the browser to discard the HTTP basic authentication credentials.

The web server stores uploaded CSV analysis source files in the analyses folder, and prefixes the original file name with the user name of the uploader, followed by a " § ". This way, the ownership of an analysis is stored as part of the file name, avoiding the need for storing additional state, separately from the analysis sources. Analysis results are not persisted, and if a user downloads or views an analysis, the results are calculated from the source, instead. However, a simple caching mechanism is employed: When an analysis is viewed for the first time, the system instance is stored by the server, so that subsequent views do not require the analysis to be re-run.

There are four Jinja2 templates used by the server. template/main. html is always rendered, as contains the necessary <script> elements as well as the navigation bar at the top. The other templates, analysis. html, existing. html and new. html are rendered nested inside the main template.

(5) Command Line Script

A small script is provided for performing ad-hoc analyses without using the server or any of its components: pymfa-cli. py can be run from the command line and takes two arguments: The first argument should point to a CSV source analysis file and the second argument specifies the desired output file location. Which exporter is used to write the output is automatically determined from the suffix of the second argument, either ". csv" or ". json".

The command line script only depends on the files contained in lib and their dependencies, which means that this small subset of files can be used to run analyses in a stand-alone environment, for example for the purpose of performing batch analyses or periodical analyses governed by a scheduler such as cron.

(6) Testing and Code Base

A unit test suite is contained in the test folder, with test data contained intest/testdata. The tests ensure the correct behavior of the CSVImporter. For this purpose, 9 tests are performed against invalid source analysis files, checking whether the importer throws a parser exception containing the correct error message. Another 4 tests check for the correct parsing of 6 different CSV dialects and different file encodings. The source analysis files for these checks implement all possible features (such as multiple inflows or conversions and delays). The script pymfa-runtests. py can be used to run the tests.

To give a rough impression of the size of the project and its sub-components, here are the lines of code (not counting empty lines and comment lines): The core implementation for performing analyses has only 148 lines of code and the command line script adds another 35 lines. The importer has 149 lines and the exporter has 80 lines. The Javascript visualization is by far the largest code component, comprising 580 lines of code with 36 lines of helper code for the navigational components of the web site. There exist an additional 356 lines of HTML template code and 178 lines of CSS. The test suite has 72 lines and there exist 20 test files of varying length.

Glossary

Anthropocene: Defines earth's most recent geologic time period as human influenced. Holds that there is global evidence that earth system processes are altered by humans. The anthropocene is the current geological epoch, viewed as the period during which human activity has been the dominant influence on our earth.

Anthropogenic: derived from human activities.

Anthropogenic circularity: Anthropogenic circularity is the human activity to enclose the material circular utilization, which is mainly consisting of reuse, remanufacturing, recycling, and recovery.

Anthropogenic metabolism: Describes the material and energy turnover of human society. It emerges from the application of systems thinking to the industrial and other man-made activities and it is a central concept of sustainable development. In modern societies, the bulk of anthropogenic material flows is related to one of the following activities: sanitation, transportation, habitation, and communication, which were "of little metabolic significance in prehistoric times".

Anthroposphere: (sometimes also referred as technosphere) is one of the earth spheres, it's the part of the environment that is made or modified by humans for use in human activities and human habitats.

Carbon neutrality: Carbon neutrality is a state of net-zero carbon dioxide emissions.

Circular economy: A circular economy is an economic model designed to minimize resource input, as well as waste and emission production.

Criticality: Criticality is the quality, state, or degree of being of the highest importance.

Dynamic MFA: is primarily used to investigate the stock buildup of materials in society (i. e. secondary resources) and in the environment (i. e. dissipative losses) based on the investigation of material flows over time.

Ecosystems: are communities of organisms that interact with each other and the abiotic environment.

Emergy analysis: Emergy analysis is a type of embodied energy analysis that can provide common units for comparison of environmental and economic goods by summing the energy of one type required directly or indirectly for production of goods.

Energy efficiency: Using less energy to accomplish a given task by using new technology, for example.

Environmental sustainability: The ability to meet humanity's current needs without compromising the ability of future generations to meet their needs.

Green chemistry: A subdiscipline of chemistry in which commercially important chemical processes are redesigned to significantly reduce environmental harm.

Industrial ecology: Defined as a field of study focused on the stages of the production processes of goods and services from a point of view of nature, trying to mimic a natural system by conserving and reusing resources.

Material flow analysis (MFA): is a systematic assessment of the flows and stocks of materials within a system defined in space and time.

Material footprint (MF): Defined as the attribution of global material extraction to domestic final demand of a country. The total material footprint is the sum of the material footprint for biomass, fossil fuels, metal ores and nonmetal ores.

Recyclability: Recyclability is the theoretical probability of an item's being recycled.

Recycling: Defined as the process of collecting and processing materials that would otherwise be thrown away as trash and turning them into new products. Recycling can benefit your community and the environment.

Recycled content: It means the proportion, by mass, of post-consumer recycled material in a product or packaging excluding any pre-consumer waste.

Resource efficiency: It means using the earth's limited resources in a sustainable manner while minimizing impacts on the environment. It allows us to create more with less and to deliver greater value with less input.

Social Metabolism: Defined as the particular form in which societies establish and maintain their material input from and output to nature; the mode in which they organize the exchange of matter and energy with their natural environment.

Substance flow analysis (SFA): Can be considered as a special type of MFA, a substance is defined as a single type of matter consisting of uniform units. If the units are atoms, the substance is called an element, such as carbon or iron; if they are molecules, it is called a chemical compound, such as carbon dioxide or iron chloride.

Sustainable consumption: The use of goods and services that satisfy basic human needs and improve the quality of life but that also minimize resource use.

Static MFA: is concerned with generating a better understanding of a material system based on simple accounting principles (i. e. mass balance equations), also provides a valuable snapshot of a system in time and is done at different levels of sophistication to investigate the patterns of material use and losses in a system.

Symbiosis: An intimate relationships or association between members of two or more species; includes mutualism, commensalism, and parasitism.

System: A system is an entity that is comprised of its components, that can be impacted by the environment, has characteristic relations and interactions between its components, and has system-specific characteristics and capacities that stem from the system processes.

Technosphere: Technological systems with own networked agency that impact the earth system; it's the part of the environment that is made or modified by humans for use in human activities and human habitats.

Life cycle assessment: Defined as a cradle-to-grave or cradle-to-cradle analysis technique to assess environmental impacts associated with all the stages of a product's life, which is from raw material extraction through materials processing, manufacture, distribution, and use.

Urban metabolism: Defined as "the sum total of the technical and socio-economic processes that occur in cities, resulting in growth, production of energy, and elimination of waste.

Urban mining: Refers to the exploitation of anthropogenic stocks, nowadays this term is widely used for describing almost any sort of material recycling. Urban mining is the process of reclaiming raw materials from waste products sent to landfill.

Urbanization: A process whereby people move from rural areas to densely populated cities.

Well-being: Is the functional integrity of the system, or in other words, the integrity of the critical system processes, that allows the system to continue its existence and realize its system-specific characteristics and capacities.

References

[1] Crutzen P J. Geology of mankind[J]. Nature, 2002, 415: 23-23.

[2] Donahue C J. The anthroposphere, material flow analysis, and chemical education [J]. Journal of Chemical Education, 2015, 92: 598-600.

[3] UNEP. Global Resources Outlook 2019: Natural Resources for the Future We Want[M]. Nairobi, Kenya: United Nations Environment Programme, 2019.

[4] Kerr R A. The coming copper peak[J]. Science, 2014, 343: 722-724.

[5] UNEP, ISWA. Global Waste Management Outlook[M]. Nairobi: UNEP, 2015.

[6] Hoornweg D, Bhada-Tata P, Kennedy C. Waste production must peak this century [J]. Nature, 2013, 502: 615-617.

[7] Hodges C A. Mineral resources, environmental issues, and land use[J]. Science, 1995, 268: 1305-1312.

[8] Liang Y, Song Q, Wu N, et al. Repercussions of COVID-19 pandemic on solid waste generation and management strategies[J]. Frontiers of Environmental Science & Engineering, 2021, 15: 115.

[9] Klinglmair M, Fellner J. Urban mining in times of raw material shortage[J]. Journal of Industrial Ecology, 2010, 14: 666-679.

[10] Ayres R U, Ayres L. A Handbook of Industrial Ecology[M]. Cheltenham: Edward Elgar Publishing, 2002.

[11] Zhang Y, Yang Z, Yu X. Urban metabolism: a review of current knowledge and directions for future study[J]. Environmental Science & Technology, 2015, 49: 11247-11263.

[12] Brunner P H, Rechberger H. Handbook of Material Flow Analysis: For Environmental, Resource, and Waste Engineers[M]. 2nd ed. Boca Raton, FL: CRC Press 2016.

[13] Baccini P, Brunner P H. Metabolism of the Anthroposphere: Analysis, Evaluation, Design[M]. Cambridge: MIT Press, 2012.

[14] Du X, Graedel T E. Global in-use stocks of the rare earth elements: a first estimate[J]. Environmental Science & Technology, 2011, 45: 4096-4101.

[15] Schimel D S. Terrestrial ecosystems and the carbon cycle[J]. Global Change Biology, 1995, 1: 77-91.

[16] Guinee J B, van den Bergh J C J M, Boelens J, et al. Evaluation of risks of metal flows and accumulation in economy and environment[J]. Ecological Economics, 1999, 30: 47-65.

[17] Graedel T, Allenby B R. Industrial Ecology and Sustainable Engineering[M]. Upper Saddle River: Prentice Hall, 2010.

[18] Graedel T E. The evolution of industrial ecology[J]. Environmental Science & Technology, 2000, 34: 28A-31A.

[19] Weisz H, Suh S, Graedel T E. Industrial ecology: the role of manufactured capital in sustainability[J]. Proceedings of the National Academy of Sciences, 2015, 112: 6260-6264.

[20] Wolman A. The metabolism of cities[J]. Scientific American, 1965, 213: 178-193.

[21] Hendriks C, Obernosterer R, Müller D, et al. Material flow analysis: a tool to support environmental policy decision making. Case-studies on the city of Vienna and the Swiss lowlands[J]. Local Environment, 2000, 5(3): 311-328.

[22] Brunner P, Rechberger H. Practical Handbook of Material Flow Analysis[M]. Boca Raton, Florida: Lewis Publishers and CRC Press LLC, 2004.

[23] Jin M. Advance of cyclical chemistry[J]. Chemical World, 2014, 55: 187-192.

[24] Ciacci L, Nuss P, Reck B, et al. Metal criticality determination for Australia, the US, and the planet—comparing 2008 and 2012 results[J]. Resources, 2016, 5: 29.

[25] Liwarska-Bizukojc E, Bizukojc M, Marcinkowski A, et al. The conceptual model of an eco-industrial park based upon ecological relationships[J]. Journal of Cleaner Production, 2009, 17: 732-741.

[26] Trost B M. Atom economy—a challenge for organic synthesis: homogeneous catalysis leads the way[J]. Angewandte Chemie International Edition in English, 1995, 34: 259-281.

[27] Muller E, Hilty L M, Widmer R, et al. Modeling metal stocks and flows: a review of dynamic material flow analysis methods[J]. Environmental Science & Technology, 2014, 48: 2102-2113.

[28] Rechberger H, Cencic O, Frühwirth R. Uncertainty in material flow analysis[J]. Journal of Industrial Ecology, 2014, 18: 159-160.

[29] Cencic O, Rechberger H. Material flow analysis with software STAN[J]. Journal of Environmental Engineering Management, 2008, 18: 3-7.

[30] Gould O, Colwill J. A framework for material flow assessment in manufacturing systems[J]. Journal of Industrial and Production Engineering, 2015, 32: 55-66.

[31] Williams E. Environmental effects of information and communications technologies[J]. Nature, 2011, 479: 354-358.

[32] China-CCTV. "40 years of reform and opening up" the transformation of innovation and stronger manufacturing power[EB/OL]. (2018-12-17). http://tv.cctv.com/2018/12/16/VIDE1A8ZjCH2eUXHSY3nqYmV181216.shtml.

[33] Habib K, Parajuly K, Wenzel H. Tracking the flow of resources in electronic waste—the case of end-of-life computer hard disk drives[J]. Environmental Science & Technology, 2015, 49: 12441-12449.

[34] Shi L, Pan L, Li Z, et al. Treatment technologies and equipments for oil-

containing sludge (in Chinese)[J]. Environmental Engineering, 2015, 33: 526-529,534.

[35] Liu L, Wang J. Application effect and evaluation of oily sludge treatment technology in the seventh oil production plant of daqing oilfield[J]. Petroleum Planning & Engineering, 2016, 27: 35-37,41.

[36] Olivetti E A, Cullen J M. Toward a sustainable materials system[J]. Science, 2018, 360: 1396-1398.

[37] Zeng X, Li J. Emerging anthropogenic circularity science: principles, practices, and challenges[J]. iScience, 2021:102237.

[38] Bodar C, Spijker J, Lijzen J, et al. Risk management of hazardous substances in a circular economy[J]. Journal of Environmental Management, 2018, 212: 108-114.

[39] Fischer-Kowalski M. Society's Metabolism: The intellectual history of materials flow analysis, Part I, I 860- I 970[J]. Journal of Industrial Ecology, 1998, 2: 61-78.

[40] OECD. The Circular Economy in Cities and Regions[M]. Paris: OECD Publishing, 2020.

[41] Eurostat. Economy-Wide Material Flow Accounts and Derived Indicators: A Methodological Guide[M]. Luxembourg: European Communities, 2001.

[42] Dai T, Liu R, Wanjun W. Material metabolism in Beijing by material flow analysis[J]. Acta Scientiae Circumstantiae, 2017, 37: 3220-3228.

[43] Remmen A, Jensen A A, Frydendal J. Life Cycle Management: A Business Guide to Sustainability[M]. [S.l.]: UNEP, 2007.

[44] Ren J, Toniolo S. Life Cycle Sustainability Assessment for Decision-Making[M]. [S.l.]:Elsevier,2020.

[45] Jacob-Lopes E, Zepka L Q, Deprá M C. Sustainability Metrics and Indicators of Environmental Impact[M]. [S.l.]: Elsevie, 2021.

[46] He P, Feng H, Hu G, et al. Life cycle cost analysis for recycling high-tech minerals from waste mobile phones in China[J]. Journal of Cleaner Production, 2020, 251: 119498.

[47] Meijers A. Philosophy of Technology and Engineering Sciences[M]. North-Holland: Elsevie, 2009.

[48] Zeng X, Xiao T, Xu G, et al. Comparing the costs and benefits of virgin and urban mining[J]. Journal of Management Science and Engineering, 2022, 7: 98-106.

[49] Lehni M. Eco-efficiency: Creating More Value with Less Impact. Geneva: WBCSD, 2000.

[50] WBCSD. Eco-efficiency learning module[EB/OL]. (2006-08-24) https://www.wbcsd.org/contentwbc/download/2328/29257/1.

[51] Eggels P G, Ansems A M M, Van Der Ven B L, TNO-report (TNO-MEP-R

2000/119)[R]. 2001.

[52] Rechberger H, Brunner P H. A new, entropy based method to support waste and resource management decisions[J]. Environmental Science & Technology, 2001, 36: 809-816.

[53] Yue Q, Lu Z W. Entropy analysis of copper products life cycle in China[J]. Resources Science, 2008, 30: 140-146.

[54] Li J, Zeng X, Chen M, et al. "Control-Alt-Delete": rebooting solutions for the e-waste problem[J]. Environmental Science & Technology, 2015, 49: 7095-7108.

[55] Li J, Zeng X, Stevels A. Ecodesign in consumer electronics: past, present and future[J]. Critical Reviews in Environmental Science and Technology, 2015, 45: 840-860.

[56] Cucchiella F, D'Adamo I, Lenny Koh S C, et al. Recycling of WEEEs: an economic assessment of present and future e-waste streams[J]. Renewable and Sustainable Energy Reviews, 2015, 51: 263-272.

[57] Dahmus J B, Gutowski T G. What gets recycled: an information theory based model for product recycling[J]. Environmental Science & Technology, 2007, 41: 7543-7550.

[58] Graedel T E. On the future availability of the energy metals[J]. Annual Review of Materials Research, 2011, 41: 323-335.

[59] Bartl A. Moving from recycling to waste prevention: a review of barriers and enables[J]. Waste Management & Research, 2014, 32: 3-18.

[60] Graedel T. The prospects for urban mining[J]. Bridge, 2011, 41: 43-50.

[61] Cover T M, Thomas J A. Elements of Information Theory[M]. [S. l.]: John Wiley & Sons, 2012.

[62] Zeng X, Zheng L, Xie H, et al. Current status and future perspective of waste printed circuit boards recycling[J]. Procedia Environmental Sciences, 2012, 16: 590-597.

[63] Watanabe T, Hirose S, Wada H, et al. A pairwise maximum entropy model accurately describes resting-state human brain networks[J]. Nat Commun, 2013, 4: 1-10.

[64] Zeng X, Li J. Designing and examining electronic waste recycling process: methodology and case study[J]. Environmental Technology, 2017, 38(5): 652-660.

[65] Johnson J, Harper E, Lifset R, et al. Dining at the periodic table: metals concentrations as they relate to recycling[J]. Environmental Science & Technology, 2007, 41: 1759-1765.

[66] Adie G U, Balogun O E, Li J H, et al. Trends in toxic metal levels in discarded laptop printed circuit boards[J]. Advanced Materials Research, 2014, 878: 413-419.

[67] Zhang X, Guan J, Guo Y, et al. Selective desoldering separation of tin-lead alloy

for dismantling of electronic components from printed circuit boards[J]. ACS Sustainable Chemistry & Engineering, 2015, 3: 1696-1700.

[68] Hadi P, Xu M, Lin C S K, et al. Waste printed circuit board recycling techniques and product utilization[J]. Journal of Hazardous Materials, 2015, 283: 234-243.

[69] Premalatha M, Abbasi T, Abbasi S A. The generation, impact, and management of e-waste: state of the art[J]. Critical Reviews in Environmental Science and Technology, 2014, 44(13): 1577-1678.

[70] Guo M, Murphy R J. LCA data quality: sensitivity and uncertainty analysis[J]. Science of the Total Environment, 2012, 435: 230-243.

[71] Zeng X, Li J. Spent rechargeable lithium batteries in e-waste: composition and its implications[J]. Frontiers of Environmental Science & Engineering, 2014, 8: 792-796.

[72] Scruggs C E, Ortolano L, Wilson M P, et al. Effect of company size on potential for REACH compliance and selection of safer chemicals[J]. Environmental Science & Policy, 2015, 45: 79-91.

[73] Rechberger H, Brunner P H. A new, entropy based method to support waste and resource management decisions[J]. Environmental Science & Technology, 2002, 36: 809-816.

[74] Liu G, Bangs C E, Müller D B. Stock dynamics and emission pathways of the global aluminium cycle[J]. Nature Climate Change, 2013, 3: 338-342.

[75] Boehme S E, Panero M A, Muñoz G R, et al. Collaborative problem solving using an industrial ecology approach[J]. Journal of Industrial Ecology, 2009, 13: 811-829.

[76] Eckelman M J, Chertow M R. Using material flow analysis to illuminate long-term waste management solutions in Oahu, Hawaii[J]. Journal of Industrial Ecology, 2009, 13: 758-774.

[77] Ragnarsdóttir K V, Sverdrup H, Koca D. Assessing long term sustainability of global supply of natural resources and materials[J]. Sustainable Development-Energy, Engineering and Technologies-Manufacturing and Environment, 2012. 83-116.

[78] Eheliyagoda D, Li J, Geng Y, et al. The role of China's aluminum recycling on sustainable resource and emission pathways[J]. Resources Policy, 2022, 76: 102552.

[79] Zeng X, Li J. Urban mining and its resources adjustment: characteristics, sustainability, and extraction[J]. SCIENTIA SINICA Terrae, 2018, 48: 288-298.

[80] Parchomenko A, Nelen D, Gillabel J, et al. Measuring the circular economy—a multiple correspondence analysis of 63 metrics[J]. Journal of Cleaner Production, 2019, 210: 200-216.

[81] O'Connor M P, Zimmerman J B, Anastas P T, et al. A strategy for material

supply chain sustainability: enabling a circular economy in the electronics industry through green engineering[J]. ACS Sustainable Chemistry & Engineering, 2016, 4: 5879-5888.

[82] Graedel T E, Nuss P. Employing considerations of criticality in product design [J]. JOM, 2014, 66: 2360-2366.

[83] Graedel T E, Barr R, Chandler C, et al. Methodology of metal criticality determination[J]. Environmental Science & Technology, 2012, 46: 1063-1070.

[84] Knobloch V, Zimmermann T, Gößling-Reisemann S. From criticality to vulnerability of resource supply: The case of the automobile industry[J]. Resources, Conservation and Recycling, 2018, 138: 272-282.

[85] Zhou Y, Li J, Rechberger H, et al. Dynamic criticality of by-products used in thin-film photovoltaic technologies by 2050[J]. Journal of Cleaner Production, 2020, 263: 121599.

[86] Colon C, Hallegatte S, Rozenberg J. Criticality analysis of a country's transport network via an agent-based supply chain model[J]. Nature Sustainability, 2021, 4: 209-215.

[87] Wang A, Wang G, Deng X, et al. Security and managment of China's critical mineral resources in the new era[J]. Science Foundation in China, 2019, 133-140.

[88] Hou Z, Chen J, Zhai M. Current status and frontiers of research on critical mineral resources[J]. Chinese Science Bulletin, 2020, 65: 3561-3652.

[89] Eheliyagoda D, Zeng X, Li J. A method to assess national metal criticality: the environment as a foremost measurement[J]. Humanities and Social Sciences Communications, 2020, 7: 43.

[90] Yan W, Wang Z, Cao H, et al. Criticality assessment of metal resources in China [J]. iScience, 2021, 24: 102524.

[91] Babbitt C W, Althaf S, Cruz Rios F, et al. The role of design in circular economy solutions for critical materials[J]. One Earth, 2021, 4: 353-362.

[92] Rademaker J H, Kleijn R, Yang Y. Recycling as a strategy against rare earth element criticality: a systemic evaluation of the potential yield of NdFeB magnet recycling[J]. Environmental Science & Technology, 2013, 47: 10129-10136.

[93] Velenturf A P M, Purnell P, Jensen P D. Reducing material criticality through circular business models: challenges in renewable energy[J]. One Earth, 2021, 4: 350-352.

[94] Tercero Espinoza L, Schrijvers D, Chen W-Q, et al. Greater circularity leads to lower criticality, and other links between criticality and the circular economy[J]. Resources, Conservation and Recycling, 2020, 159: 104718.

[95] Das S. Achieving carbon neutrality in the global aluminum industry[J]. JOM, 2012, 64: 285-290.

[96] Raabe D, Tasan C C, Olivetti E A. Strategies for improving the sustainability of

structural metals[J]. Nature, 2019, 575: 64-74.

[97] Chen J M. Carbon neutrality: toward a sustainable future[J]. The Innovation, 2021, 2: 100127.

[98] Iranpour R, Stenstrom M, Tchobanoglous G, et al. Environmental engineering: energy value of replacing waste disposal with resource recovery [J]. Science, 1999, 285: 706-711.

[99] Lu X, Zhang Z, Hiraki T, et al. A solid-state electrolysis process for upcycling aluminium scrap[J]. Nature, 2022, 606: 511-515.

[100] Liu S. Analysis of electronic waste recycling in the United States and Potential Application in China[J]. New York: Columbia University, 2014.

[101] USEPA. Electronics Environmental Benefits Calculator [spreadsheet]. FEC Publications and Resources- Life Cycle Assessment Data for Multi-Function Devices & Printers [EB/OL]. (2019-04-14). https://www.epa.gov/sites/production/files/2018-02/documents/lca_mfd_printer.pdf.

[102] Norgate T, Haque N. Energy and greenhouse gas impacts of mining and mineral processing operations[J]. Journal of Cleaner Production, 2010, 18: 266-274.

[103] Ashby M F, Johnson K. Materials and design: the art and science of material selection in product design[M]. 3rd ed. Waltham, USA: Butterworth-Heinemann, 2014.

[104] Wang P, Wang H, Chen W Q, et al. Carbon neutrality needs a circular metal-energy nexus[J]. Fundamental Research, 2022, 2: 392-395.

[105] Nuss P, Eckelman M J. Life cycle assessment of metals: a scientific synthesis [J]. PloS one, 2014, 9: e101298.

[106] Klotz M, Haupt M, Hellweg S. Limited utilization options for secondary plastics may restrict their circularity[J]. Waste Management, 2022, 141: 251-270.

[107] Spatari S, Bertram M, Fuse K, et al. The contemporary European copper cycle: 1 year stocks and flows[J]. Ecological Economics, 2002, 42: 27-42.

[108] Rechberger H, Graedel T E. The contemporary European copper cycle: statistical entropy analysis[J]. Ecological Economics, 2002, 42: 59-72.

[109] Ciacci L, Vassura I, Passarini F. Urban mines of copper: size and potential for recycling in the EU[J]. Resources, 2017, 6(1).

[110] Soulier M, Glöser-Chahoud S, Goldmann D, et al. Dynamic analysis of European copper flows[J]. Resources, Conservation and Recycling, 2018, 129: 143-152.

[111] Graedel T E, van Beers D, Bertram M, et al. Multilevel cycle of anthropogenic copper[J]. Environmental Science & Technology, 2004, 38: 1242-1252.

[112] Zhang L, Yang J, Cai Z, et al. Analysis of copper flows in China from 1975 to 2010[J]. Science of the Total Environment, 2014, 478: 80-89.

[113] Kapur A, Bertram M, Spatari S, et al. The contemporary copper cycle of Asia [J]. Journal of Material Cycles and Waste Management, 2003, 5: 143-156.

[114] Daigo I, Hashimoto S, Matsuno Y, et al. Material stocks and flows accounting

for copper and copper-based alloys in Japan[J]. Resources, Conservation and Recycling, 2009, 53: 208-217.

[115] Wang M, Chen W, Li X. Substance flow analysis of copper in production stage in the U. S. from 1974 to 2012[J]. Resources, Conservation and Recycling, 2015, 105: 36-48.

[116] Chen W, Wang M, Li X. Analysis of copper flows in the United States: 1975—2012[J]. Resources, Conservation and Recycling, 2016, 111: 67-76.

[117] Spatari S, Bertram M, Gordon R B, et al. Twentieth century copper stocks and flows in North America: a dynamic analysis[J]. Ecological Economics, 2005, 54: 37-51.

Book Review

Material Flow Analysis is a key tool to understand, design, and improve the anthroposphere. The well-illustrated textbook by Xianlai Zeng is a comprehensive representation of the state of the art. It shows that China is among the leaders developing and promoting this novel scientific methodology and its application. Besides introducing the basics of MFA, the book covers a wealth of approaches, examples, and valuable data that is useful for students, professionals and scholars in the fields of civil and environmental engineering, urban planning, and industrial design. It will be highly instrumental in establishing tomorrow's sustainable societies.

—Paul H. Brunner, global father of material flow analysis methodology and professor emeritus at Vienna University of Technology

物质流分析是理解、设计和促进人类圈的关键工具，曾现来精心编写的物质流分析教材是该领域最新发展的代表。中国近年来在发展和促进物质流创新方法及应用方面，发挥了领导及先锋的作用。除了介绍物质流分析的基础，该教材涵盖了大量的方法、案例及数据，对土木与环境工程、城市规划、工业设计领域的学生、专业人员和学者很有帮助。总之，该教材对未来建设可持续社会具有指导价值。

——保罗·汉斯·布鲁纳，全球物质流分析方法学之父，维也纳技术大学荣休教授

The strategic goal of sustainable development in the Anthropocene is to decouple economic and social development from resource and environmental consumption, and to improve human well-being within the earth's planetary boundaries. The scientific tool is material flow analysis, aiming to minimize the life-cycle material flows at various scales with effectiveness. It is a great pleasure to see that Dr. Xianlai Zeng has led the writing and publication of this English textbook that combines high theoretical value with practical usefulness.

—Dajian Zhu, distinguished professor, director of Institute of Governance for Sustainability, Tongji University

人类世可持续发展的战略目标是经济社会发展与资源环境消耗脱钩，提高地球行星边界内的人类福祉。科学工具是物质流分析，以便在各个尺度有效地实现全寿命周期物质流的最小化。很高兴看到曾现来博士主持撰写出版了这本理论高度和实用价值兼具的英文教科书。

——诸大建(同济大学特聘教授，可持续发展与管理研究所所长)

Tsinghua School of Environment is one of the earliest organizations in initialing material flow analysis in China. Material flow analysis is playing an increasing role in identifying the environmental problem, which is gradually recognized in recent two decades. Dr. Xianlai Zeng is writing *Material Flow Analysis and Its Applications* as the first English textbook in this area in China. It not only systematically addresses the method and procedure, but also contains plenty of case studies, which can satisfy for decision making of various environmental research areas.

—Yi Liu, professor and dean of School of Environment, Tsinghua University

清华大学环境学院是国内较早开展物质流分析研究的单位之一,物质流分析作为识别环境问题的重要工具,20多年来逐渐得到大家的认同。曾现来主持编写的《物质流分析及其应用》是国内第一部该领域的英文教材,系统阐述物质流分析方法,并且具有丰富的案例,可以满足环境领域不同方向决策分析的需要。

——刘毅,清华大学环境学院教授

Matter flow analysis is an important content of industrial ecology and a crucially fundamental method for studying industrial ecology problems, and has been widely adopted in academic research in environmental science&engineering and recycling science&engineering. The English textbook *Material Flow Analysis and Its Applications* written by Xianlai Zeng fills the gap in the study of this method in English in China. The book systematically addresses the theory of material flow and introduces many latest cases and new methods, which is highly academic, practical, and instructive, and is believed to be of great benefit to the future application of material flow analysis.

—Meiting Ju, professor of School of Environmental Science and Engineering, Nankai University; associate director, Steering Committee for Environmental Science and Engineering Teaching in Higher Education Institutions, Ministry of Education

物质流分析是产业生态学重要内容,也是研究产业生态学问题的关键基础方法,目前已在环境科学与工程、资源循环科学与工程等专业的学术研究中得到广泛应用。曾现来编写的《物质流分析及其应用》英文教材填补了国内英文学习该方法的空白,书中系统阐述了物质流理论,并介绍了大量的最新案例及新方法,具有很强的学术性、实践性和指导性,相信会对物质流分析的未来广泛应用大有裨益。

——鞠美庭,南开大学环境科学与工程学院教授,教育部高等学校环境科学与工程教学指导委员会副主任委员

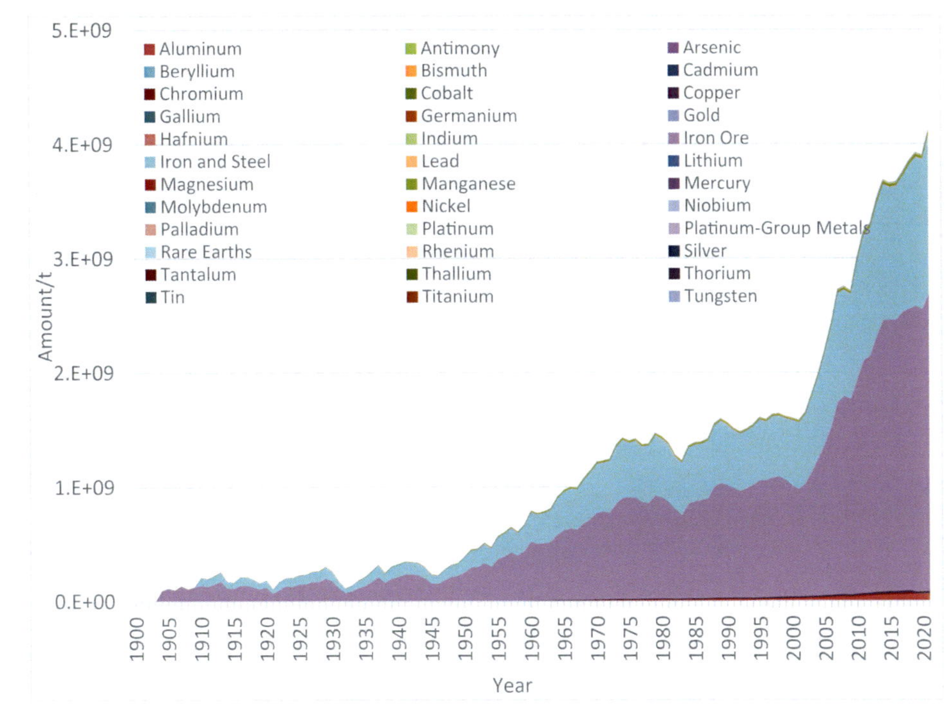

Figure 1 - 1　Global yearly primary mining production of metal mineral commodity (Data source from USGS)

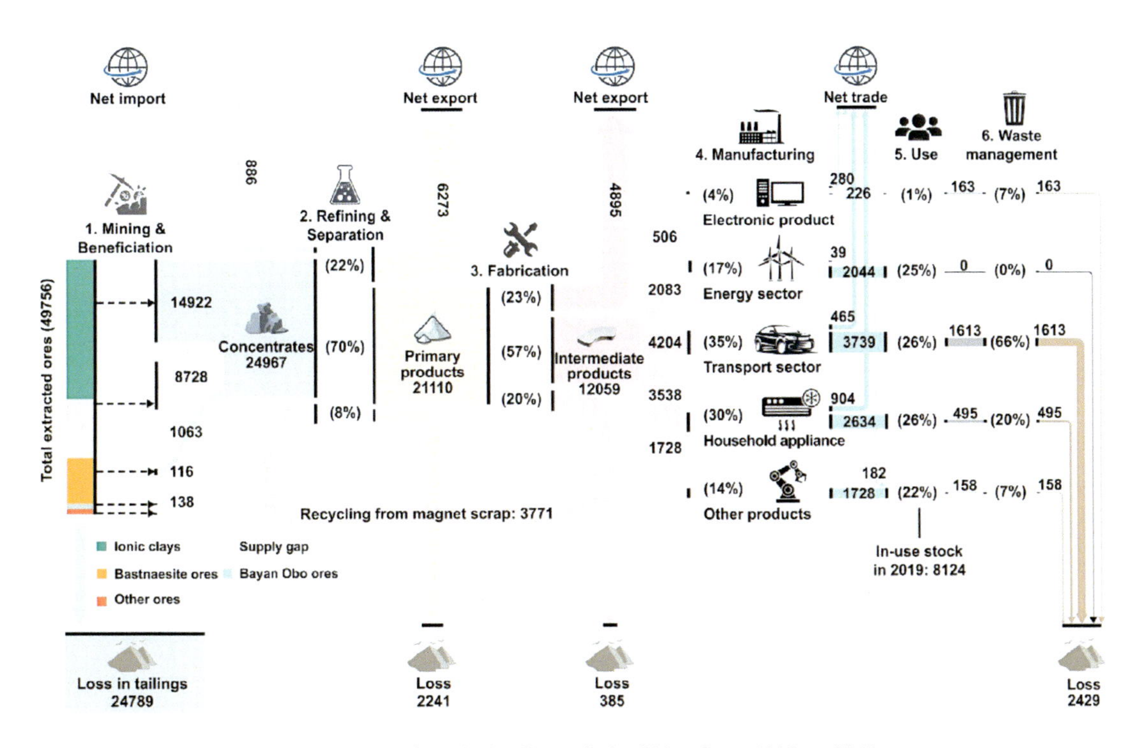

Figure 3 - 4　Cumulative Dy cycle in China from 1990 to 2019

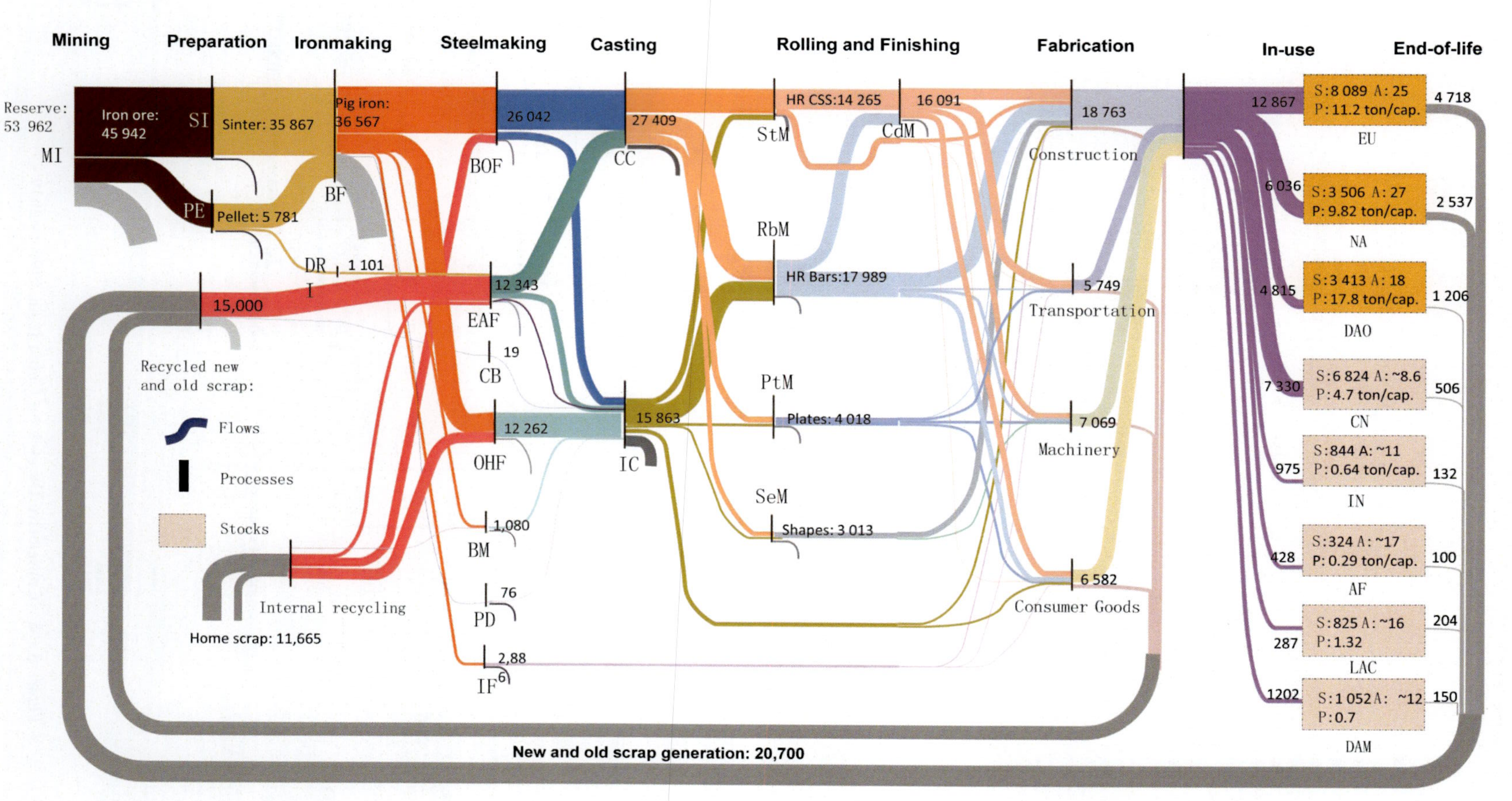

Period: 1900—2015; Unit: million tons; S: Stock amount (Unit: million tons); A: Average age of existing stock (unit: year); P: Stock per capita (cap. in box)

Figure 3–1 The global historical steel cycle in a Sankey diagram where the numbers represent the accumulated annual flows over the past 115 years

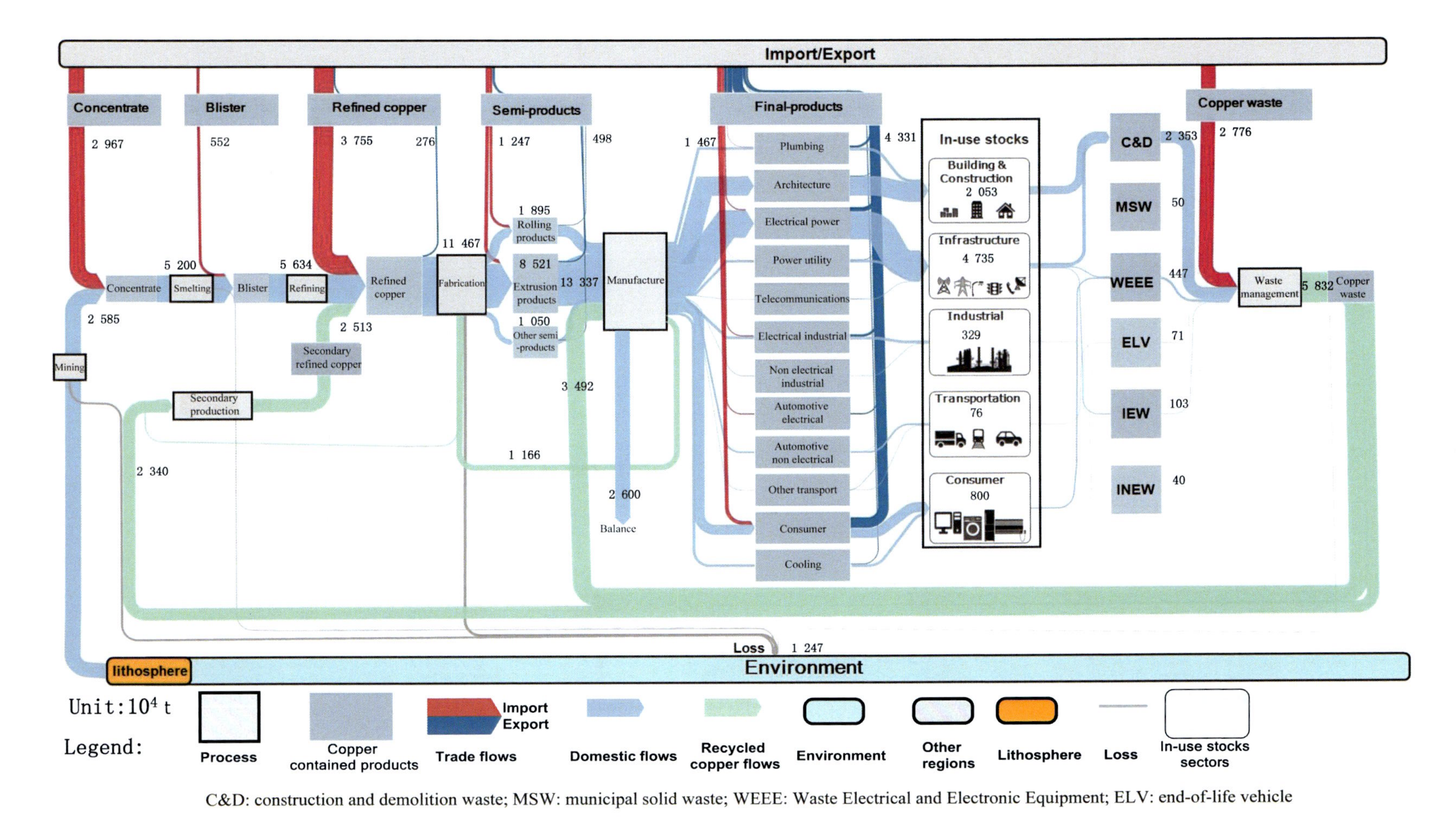

C&D: construction and demolition waste; MSW: municipal solid waste; WEEE: Waste Electrical and Electronic Equipment; ELV: end-of-life vehicle

Figure 3–2 Historical cumulative copper stocks and flows from 1950 to 2015

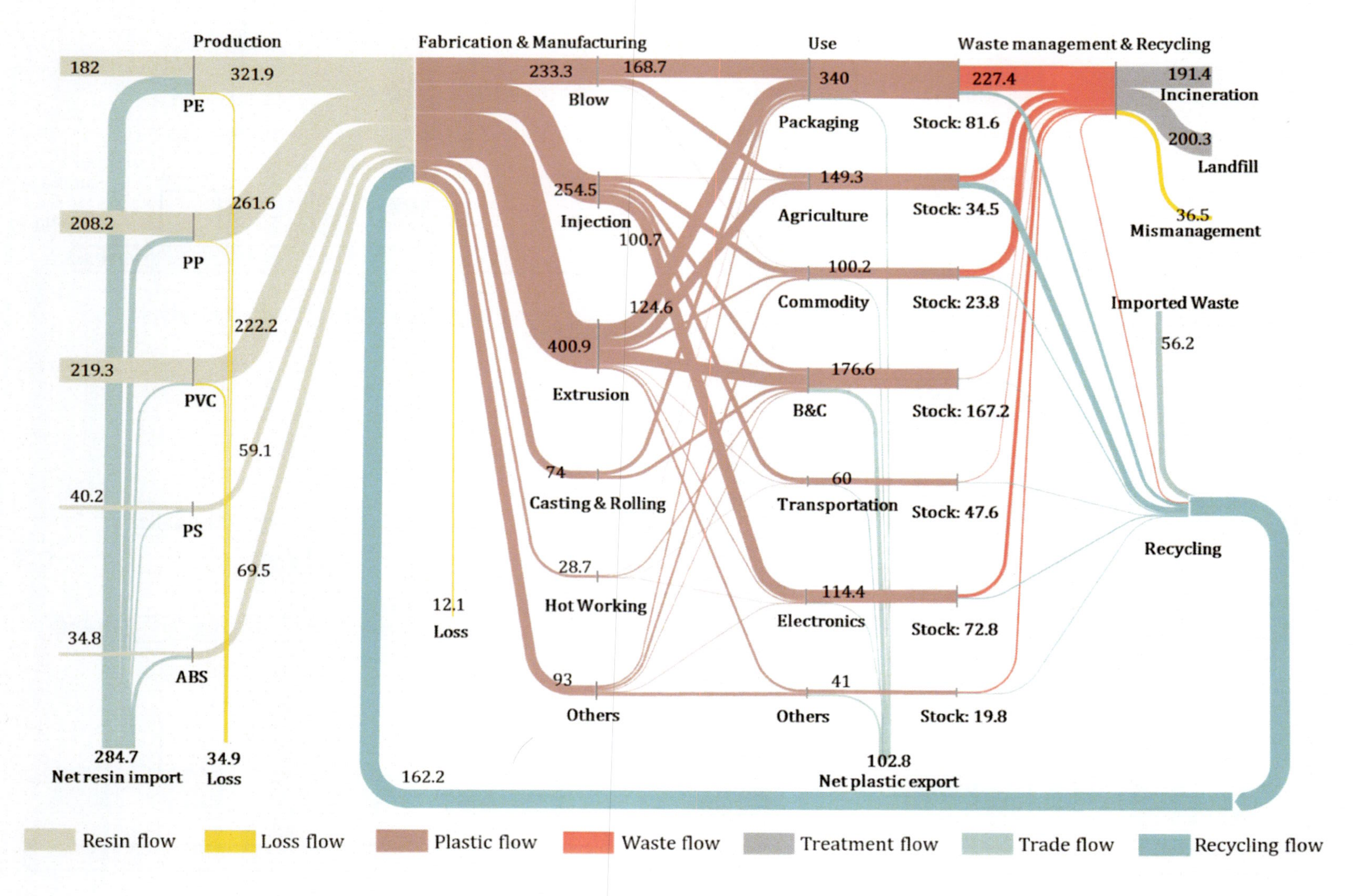

Figure 3-6　China's five commodity plastics material flow analysis from 2000 to 2019

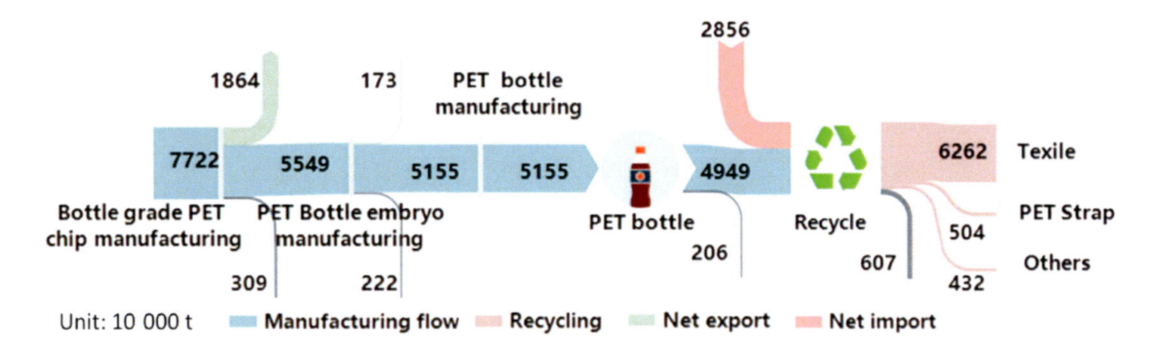

Figure 3 - 7 Material flow analysis of PET bottles in China from 2000 to 2018

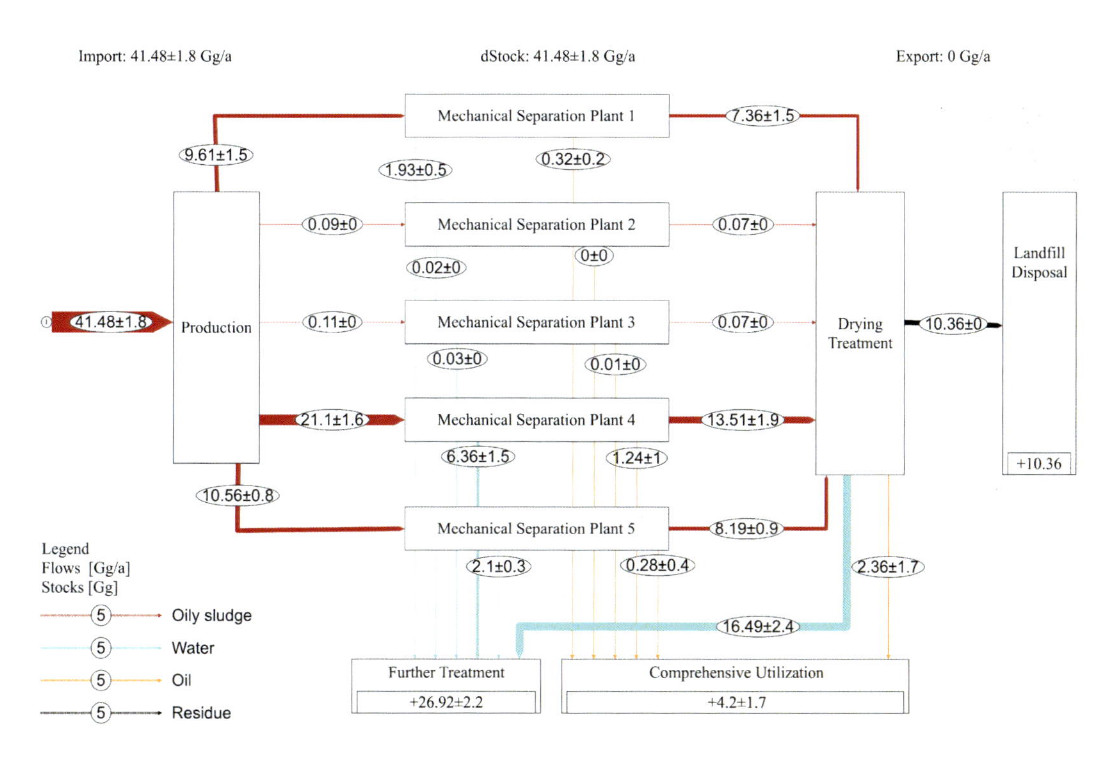

Figure 3 - 13 Material flow of oily sludge mechanical separation treatment in China

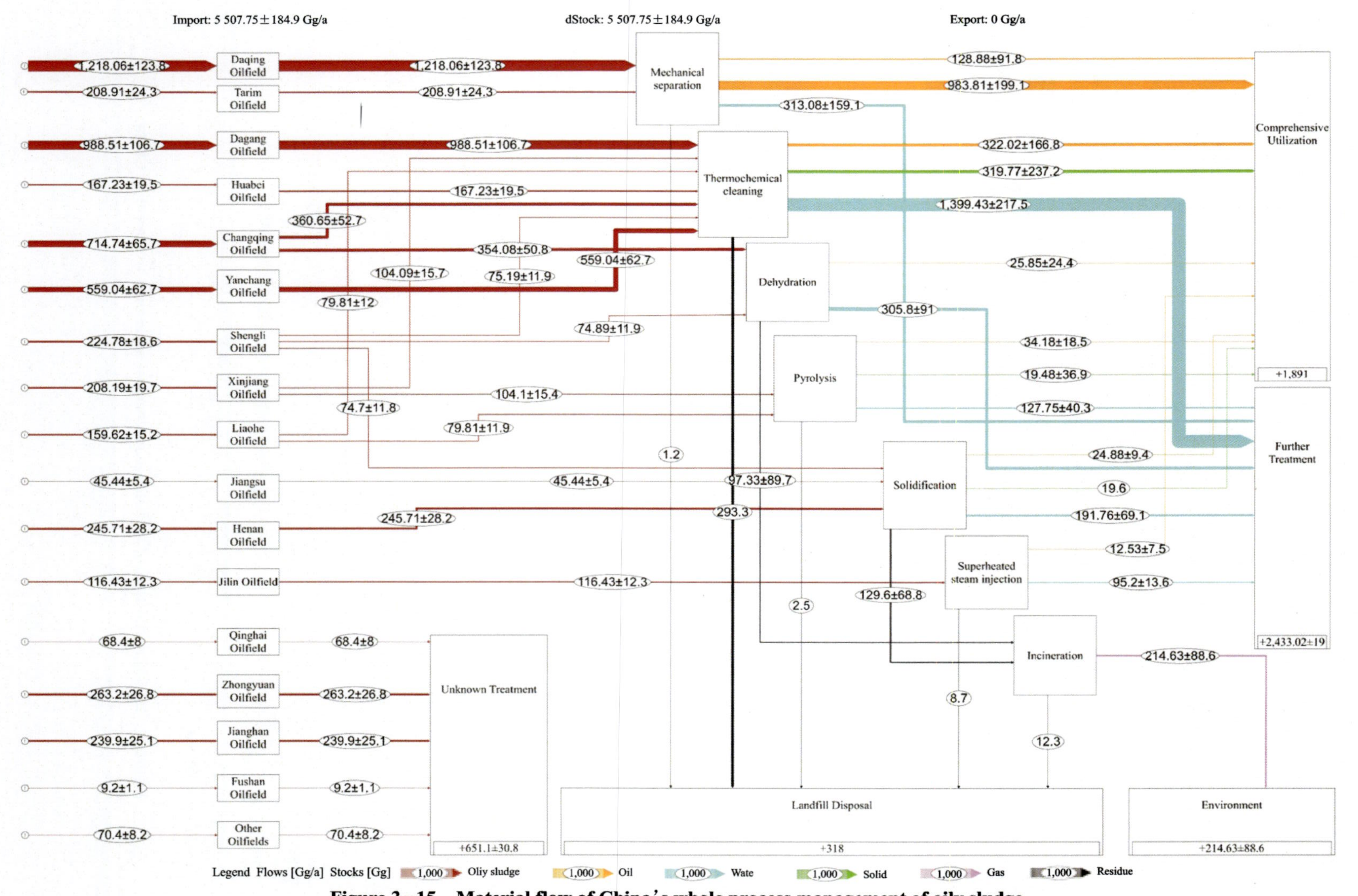

Figure 3-15 Material flow of China's whole process management of oily sludge

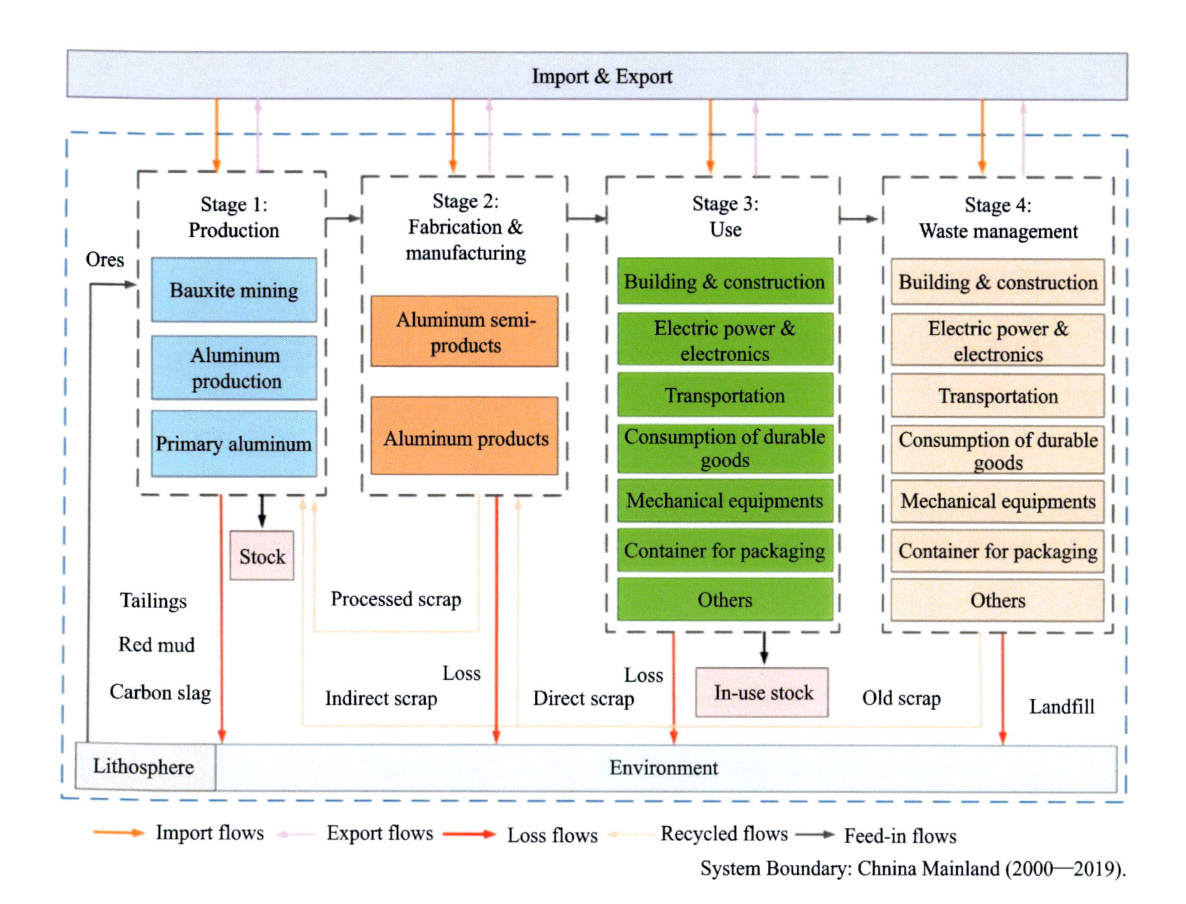

Figure 4 - 16 STAF framework applied to the aluminum life cycle

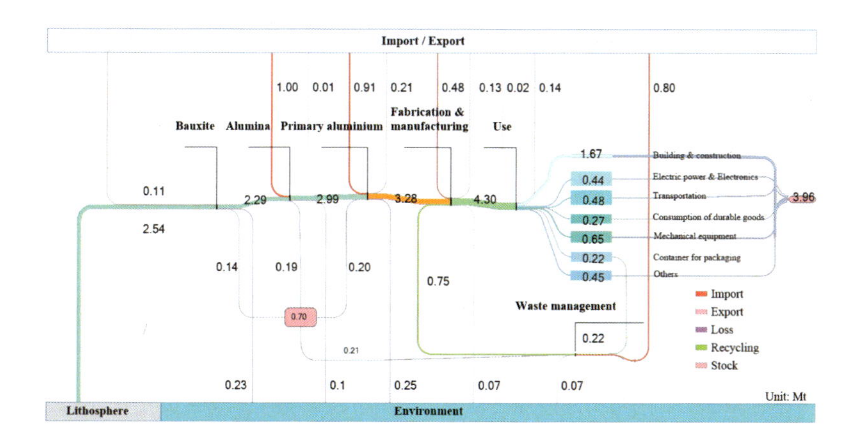

Figure 4 - 19 Life cycle-based aluminum flows in China for 2000, 2005, 2010, 2015 and 2019

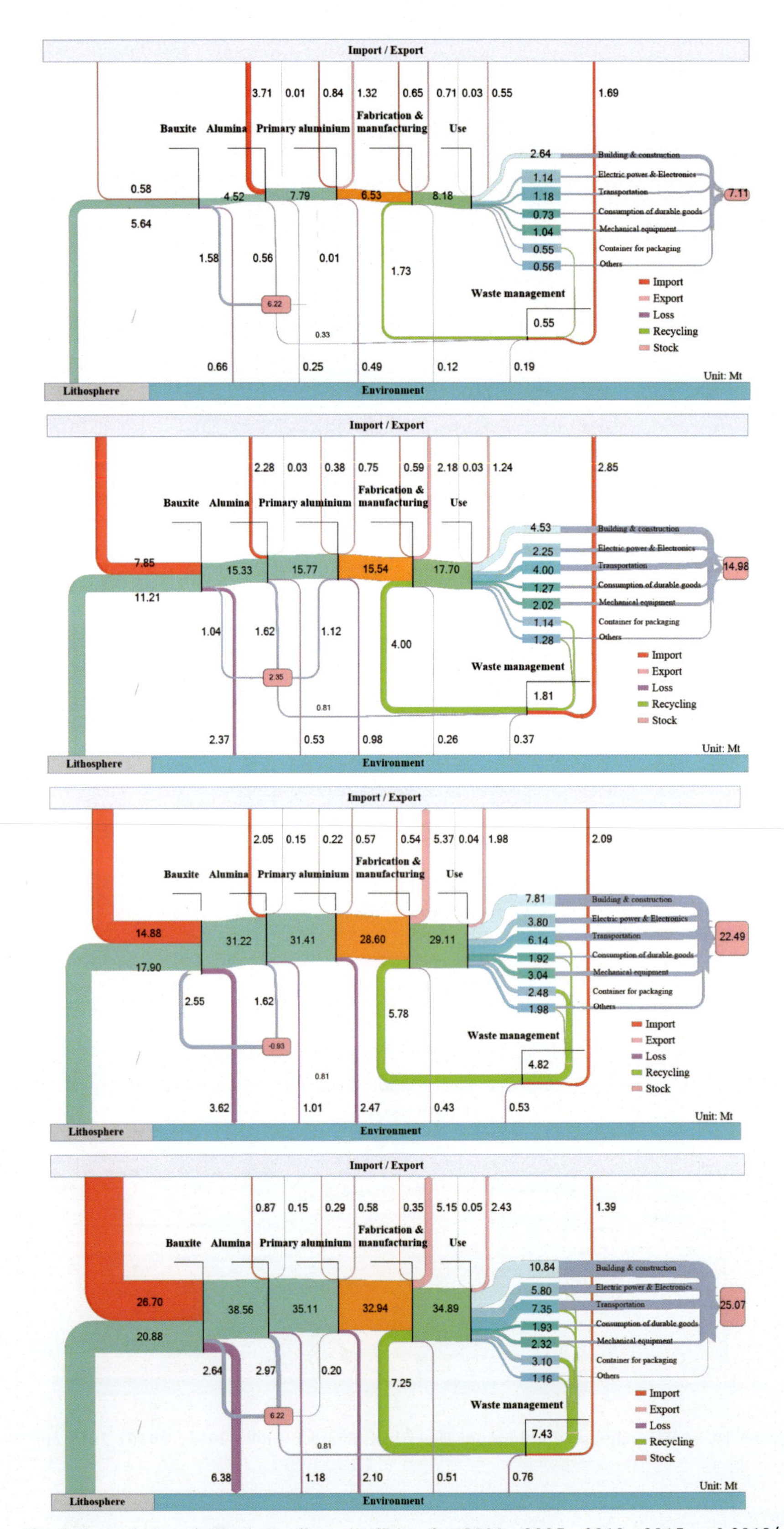

Figure 4－19 Life cycle-based aluminum flows in China for 2000, 2005, 2010, 2015 and 2019(continued)